Mysticism:
Christian and Buddhist

Mysticism: Christian and Buddhist

The Eastern and Western Way

D. T. Suzuki

PERENNIAL LIBRARY
Harper & Row, Publishers
New York, Evanston, San Francisco, London

This volume reprints Volume XII of the World Perspectives Series which is planned and edited by Ruth Nanda Anshen. Dr. Anshen's Epilogue appears on p. 235.

MYSTICISM: CHRISTIAN AND BUDDHIST

Copyright © 1957 by Daisetz Teitaro Suzuki. All rights reserved. Printed in the United States of America. No part of this book may be used or reproduced in any manner without written permission except in the case of brief quotations embodied in critical articles and reviews. For information address Harper & Row, Publishers, Inc., 49 East 33rd Street, New York, N.Y. 10016. Published simultaneously in Canada by Fitzhenry and Whiteside Limited, Toronto.

First PERENNIAL LIBRARY edition published 1971.

STANDARD BOOK NUMBER: 06–080218–9

Contents

Preface   vii

SECTION ONE

1. Meister Eckhart and Buddhism   3
2. The Basis of Buddhist Philosophy   39
3. "A Little Point" and Satori   85
4. Living in the Light of Eternity   105

APPENDICES

5. Transmigration   129
6. Crucifixion and Enlightenment   145

SECTION TWO

7. Kono-mama ("I Am That I Am")   159

APPENDICES

8. Notes on "Namu-amida-butsu"   179
9. Rennyo's Letters   185
10. From Saichi's Journals   193

Epilogue   235

Preface

This book has no pretension to be a thorough, systematic study of the subject. It is more or less a collection of studies the author has written from time to time in the course of his readings, especially of Meister Eckhart as representative of Christian mysticism. For Eckhart's thoughts come most closely to those of Zen and Shin. Zen and Shin superficially differ: one is known as *Jiriki*, the "self-power" school, while the other is *Tariki*, the "other-power" school. But there is something common to both, which will be felt by the reader. Eckhart, Zen, and Shin thus can be grouped together as belonging to the great school of mysticism. The underlying chain of relationship among the three may not be always obvious in the following pages. The author's hope, however, is that they are provocative enough to induce Western scholars to take up the subject for their study.

The author wishes to acknowledge his debts to the two English translations of Meister Eckhart, the first by C. de B. Evans and the second by Raymond B. Blakney, from which he has very liberally quoted.

DAISETZ T. SUZUKI

New York, 1957

Mysticism:
Christian and Buddhist

Section One

chapter 1

Meister Eckhart[1] and Buddhism

I

In the following pages I attempt to call the reader's attention to the closeness of Meister Eckhart's way of thinking to that of Mahāyāna Buddhism, especially of Zen Buddhism. The attempt is only a tentative and sketchy one, far from being systematic and exhaustive. But I hope the reader will find something in it which evokes his curiosity enough to undertake further studies of this fascinating topic.

When I first read—which was more than a half century ago—a little book containing a few of Meister Eckhart's sermons, they impressed me profoundly, for I never expected that any Christian thinker ancient or modern could or would cherish such daring thoughts as expressed in those sermons. While I do not remember which sermons made up

---

[1] There are two English translations of Eckhart, one British and the other American. The British, in two volumes, is by C. de B. Evans, published by John M. Watkins, London, 1924. The American translation is by Raymond B. Blakney, published by Harper & Brothers, New York, 1941. Neither of them is a complete translation of all of Eckhart's known works in German. Franz Pfeiffer published in 1857 a collection of Eckhart's works, chiefly in the High German dialect of Strassburg of the fourteenth century. This edition was reprinted in 1914. Blakney's and Evans' translations are mainly based on the Pfeiffer edition. In the present book, "Blakney" refers to the Blakney translation and "Evans" to the Evans, Vol. I, while "Pfeiffer" means his German edition of 1914.

the contents of the little book, the ideas expounded there closely approached Buddhist thoughts, so closely indeed, that one could stamp them almost definitely as coming out of Buddhist speculations. As far as I can judge, Eckhart seems to be an extraordinary "Christian."

While refraining from going into details we can say at least this: Eckhart's Christianity is unique and has many points which make us hesitate to classify him as belonging to the type we generally associate with rationalized modernism or with conservative traditionalism. He stands on his own experiences which emerged from a rich, deep, religious personality. He attempts to reconcile them with the historical type of Christianity modeled after legends and mythology. He tries to give an "esoteric" or inner meaning to them, and by so doing he enters fields which were not touched by most of his historical predecessors.

First, let me give you the views Eckhart has on time and creation. These are treated in his sermon delivered on the commemoration day for St. Germaine. He quotes a sentence from Ecclesiasticus: "In his days he pleased God and was found just." Taking up first the phrase "In his days," he interprets it according to his own understanding:

> ... there are more days than one. There is the soul's day and God's day. A day, whether six or seven ago, or more than six thousand years ago, is just as near to the present as yesterday. Why? Because all time is contained in the present Now-moment. Time comes of the revolution of the heavens and day began with the first revolu-

tion. The soul's day falls within this time and consists of the natural light in which things are seen. God's day, however, is the complete day, comprising both day and night. It is the real Now-moment, which for the soul is eternity's day, on which the Father begets his only begotten Son and the soul is reborn in God.[2]

The soul's day and God's day are different. In her natural day the soul knows all things above time and place; nothing is far or near. And that is why I say, this day all things are of equal rank. To talk about the world as being made by God to-morrow, yesterday, would be talking nonsense. God makes the world and all things in this present now. Time gone a thousand years ago is now as present and as near to God as this very instant. The soul who is in this present now, in her the Father bears his one-begotten Son and in that same birth the soul is born back into God. It is one birth; as fast as she is reborn into God the Father is begetting his only Son in her.[3]

God the Father and the Son have nothing to do with time. Generation is not in time, but at the end and limit of time. In the past and future movements of things, your heart flits about; it is in vain that you attempt to know eternal things; in divine things, you should be occupied intellectually. . . .[4]

. Again, God loves for his own sake, acts for his own sake: that means that he loves for the sake of love and acts for the sake of action. It cannot be doubted that God would never have begot his Son in eternity if [his idea of] creation were other than [his act of] creation. Thus God

[2] Blakney, p. 212.
[3] Evans, p. 209.
[4] Blakney, p. 292.

created the world so that he might keep on creating. The past and future are both far from God and alien to his way.[5]

From these passages we see that the Biblical story of Creation is thoroughly contradicted; it has not even a symbolic meaning in Eckhart, and, further, his God is not at all like the God conceived by most Christians. God is not in time mathematically enumerable. His creativity is not historical, not accidental, not at all measurable. It goes on continuously without cessation with no beginning, with no end. It is not an event of yesterday or today or tomorrow, it comes out of timelessness, of nothingness, of Absolute Void. God's work is always done in an absolute present, in a timeless "now which is time and place in itself." God's work is sheer love, utterly free from all forms of chronology and teleology. The idea of God creating the world out of nothing, in an absolute present, and therefore altogether beyond the control of a serial time conception will not sound strange to Buddhist ears. Perhaps they may find it acceptable as reflecting their doctrine of Emptiness (śūnyatā).

2

Below are further quotations from Eckhart giving his views on "being," "life," "work," etc.:

Being is God. . . . God and being are the same—or God has being from another and thus himself is not God. . . . Everything that is has the fact of its being

[5] *Ibid.*, p. 62.

through being and from being. Therefore, if being is something different from God, a thing has its being from something other than God. Besides, there is nothing prior to being, because that which confers being creates and is a creator. To create is to give being out of nothing.[6]

Eckhart is quite frequently metaphysical and makes one wonder how his audience took to his sermons—an audience which is supposed to have been very unscholarly, being ignorant of Latin and all the theologies written in it. This problem of being and God's creating the world out of nothing must have puzzled them very much indeed. Even the scholars might have found Eckhart beyond their understanding, especially when we know that they were not richly equipped with the experiences which Eckhart had. Mere thinking or logical reasoning will never succeed in clearing up problems of deep religious significance. Eckhart's experiences are deeply, basically, abundantly rooted in God as Being which is at once being and not-being: he sees in the "meanest" thing among God's creatures all the glories of his is-ness (*isticheit*). The Buddhist enlightenment is nothing more than this experience of is-ness or suchness (*tathatā*), which in itself has all the possible values (*guna*) we humans can conceive.

God's characteristic is being. The philosopher says one creature is able to give another life. For in being, mere being, lies all that is at all. Being is the first name. Defect means lack of being. Our whole life ought to be being. So far as our life is being, so far it is in God. So far as our

[6] *Ibid.*, p. 278.

life is feeble but taking it as being, it excels anything life can ever boast. I have no doubt of this, that if the soul had the remotest notion of what being means she would never waver from it for an instant. The most trivial thing perceived in God, a flower for example as espied in God, would be a thing more perfect than the universe. The vilest thing present in God as being is better than angelic knowledge.[7]

This passage may sound too abstract to most readers. The sermon is said to have been given on the commemoration day of the "blessed martyrs who were slain with the swords." Eckhart begins with his ideas about death and suffering which come to an end like everything else that belongs to this world. He then proceeds to tell us that "it behooves us to emulate the dead in dispassion (*niht betrüeben*) towards good and ill and pain of every kind," and he quotes St. Gregory: "No one gets so much of God as the man who is thoroughly dead," because "death gives them [martyrs] being,—they lost their life and found their being." Eckhart's allusion to the flower as espied in God reminds us of Nansen's interview with Rikko in which the Zen master also brings out a flower in the monastery courtyard.

It is when I encounter such statements as these that I grow firmly convinced that the Christian experiences are not after all different from those of the Buddhist. Terminology is all that divides us and stirs us up to a wasteful dissipation of energy. We must however weigh the matter carefully and see

[7] Evans, p. 206.

whether there is really anything that alienates us from one another and whether there is any basis for our spiritual edification and for the advancement of a world culture.

When God made man, he put into the soul his equal, his active, everlasting masterpiece. It was so great a work that it could not be otherwise than the soul and the soul could not be otherwise than the work of God. God's nature, his being, and the Godhead all depend on his work in the soul. Blessed, blessed be God that he does work in the soul and that he loves his work! That work is love and love is God. God loves himself and his own nature, being and Godhead, and in the love he has for himself he loves all creatures, not as creatures but as God. The love God bears himself contains his love for the whole world.[8]

Eckhart's statement regarding God's self-love which "contains his love for the whole world" corresponds in a way to the Buddhist idea of universal enlightenment. When Buddha attained the enlightenment, it is recorded, he perceived that all beings non-sentient as well as sentient were already in the enlightenment itself. The idea of enlightenment may make Buddhists appear in some respects more impersonal and metaphysical than Christians. Buddhism thus may be considered more scientific and rational than Christianity which is heavily laden with all sorts of mythological paraphernalia. The movement is now therefore going on among Christians to denude the religion of this unnecessary historical appendix. While it is difficult to predict

[8] Blakney, pp. 224–5.

how far it will succeed, there are in every religion some elements which may be called irrational. They are generally connected with the human craving for love. The Buddhist doctrine of enlightenment is not after all such a cold system of metaphysics as it appears to some people. Love enters also into the enlightenment experience as one of its constituents, for otherwise it could not embrace the totality of existence. The enlightenment does not mean to run away from the world, and to sit cross-legged at the peak of the mountain, to look down calmly upon a bomb-struck mass of humanity. It has more tears than we imagine.

Thou shalt know him [God] without image, without semblance and without means.—"But for me to know God thus, with nothing between, I must be all but he, he all but me."—I say, God must be very I, I very God, so consummately one that this he and this I are one "is," in this is-ness working one work eternally; but so long as this he and this I, to wit, God and the soul, are not one single here, one single now, the I cannot work with nor be one with that he.[9]

What is life? God's being is my life, but if it is so, then what is God's must be mine and what is mine God's. God's is-ness is my is-ness, and neither more nor less. The just live eternally with God, on a par with God, neither deeper nor higher. All their work is done by God and God's by them.[10]

Going over these quotations, we feel that it was natural that orthodox Christians of his day accused

[9] Evans, p. 247.
[10] Blakney, p. 180.

Eckhart as a "heretic" and that he defended himself. Perhaps it is due to our psychological peculiarities that there are always two opposing tendencies in the human way of thinking and feeling; extrovert and introvert, outer and inner, objective and subjective, exoteric and esoteric, traditional and mystical. The opposition between these two tendencies or temperaments is often too deep and strong for any form of reconciliation. This is what makes Eckhart complain about his opponents not being able to grasp his point. He would remonstrate: "Could you see with my heart you would understand my words, but, it is true, for the truth itself has said it." [11] Augustine is however tougher than Eckhart: "What is it to me though any comprehend not this!" [12]

3

One of Eckhart's heresies was his pantheistic tendency. He seemed to put man and God on an equal footing: "The Father begets his Son in me and I am there in the same Son and not another." [13] While it is dangerous to criticize Eckhart summarily as a pantheist by picking one or two passages at random from his sermons, there is no doubt that

[11] Evans, p. 38.
[12] Quoted by Eckhart, Blakney, p. 305.
[13] *Cf.* Blakney, p. 214: "The soul that lives in the present Now-moment is the soul in which the Father begets his only begotten Son and in that birth the soul is born again into God. It is one birth, as fast as she is reborn into God the Father is betting his only Son in her." (The last sentence is from Evans, p. 209.)

his sermons contain many thoughts approaching pantheism. But unless the critics are a set of ignorant misinterpreters with perhaps an evil intention to condemn him in every way as a heretic, a fair-minded judge will notice that Eckhart everywhere in his sermons is quite careful to emphasize the distinction between the creature and the creator as in the following:

"Between the only begotten Son and the soul there is no distinction." This is true. For how could anything white be distinct from or divided from whiteness? Again, matter and form are one in being; living and working. Yet matter is not, on this account, form, or conversely. So in the proposition. A holy soul is one with God, according to John 17:21. That they all may be one in us, even as we are one. Still the creature is not the creator, nor is the just man God.[14]

God and Godhead are as different as earth is from heaven. Moreover I declare: the outward and the inward man are as different, too, as earth and heaven. God is higher, many thousand miles. Yet God comes and goes. But to resume my argument: God enjoys himself in all things. The sun sheds his light upon all creatures, and anything he sheds his beams upon absorbs them, yet he loses nothing of his brightness.[15]

From this we can see most decidedly that Eckhart was far from being a pantheist. In this respect Mahāyāna Buddhism is also frequently and erroneously stamped as pantheistic, ignoring altogether a world of particulars. Some critics seem to be ready and simple-minded enough to imagine that all doc-

[14] *Ibid.*, "The Defense," p. 303.
[15] Evans, pp. 142–3.

trines that are not transcendentally or exclusively monotheistic are pantheistic and that they are for this reason perilous to the advancement of spiritual culture.

It is true that Eckhart insists on finding something of a Godlike nature in each one of us, otherwise the birth of God's only Son in the soul would be impossible and his creatures would forever be something utterly alienated from him. As long as God is love, as creator, he can never be outside the creatures. But this cannot be understood as meaning the oneness of one with the other in every possible sense. Eckhart distinguishes between the inner man and the outer man and what one sees and hears is not the same as the other. In a sense therefore we can say that we are not living in an identical world and that the God one conceives for oneself is not at all to be subsumed under the same category as the God for another. Eckhart's God is neither transcendental nor pantheistic.

God goes and comes, he works, he is active, he becomes all the time, but Godhead remains immovable, imperturbable, inaccessible. The difference between God and Godhead is that between heaven and earth and yet Godhead cannot be himself without going out of himself, that is, he is he because he is not he. This "contradiction" is comprehended only by the inner man, and not by the outer man, because the latter sees the world through the senses and intellect and consequently fails to experience the profound depths of Godhead.

Whatever influence Eckhart might have received from the Jewish (Maimonides), Arabic (Avicenna), and Neoplatonic sources, there is no doubt that he had his original views based on his own experiences, theological and otherwise, and that they were singularly Mahāyānistic. Coomaraswamy is quite right when he says:

> Eckhart presents an astonishingly close parallel to Indian modes of thought; some whole passages and many single sentences read like a direct translation from Sanskrit. . . . It is not of course suggested that any Indian elements whatever are actually present in Eckhart's writing, though there are some Oriental factors in the European tradition, derived from neo-Platonic and Arabic sources. But what is proved by analogies is not the influence of one system of thought upon another, but the coherence of the metaphysical tradition in the world and at all times.[16]

4

It is now necessary to examine Eckhart's close kinship with Mahāyāna Buddhism and especially with Zen Buddhism in regard to the doctrine of Emptiness.

The Buddhist doctrine of Emptiness is unhappily greatly misunderstood in the West. The word "emptiness" or "void" seems to frighten people away, whereas when they use it among themselves, they do not seem to object to it. While some Indian thought is described as nihilistic, Eckhart has never been accused of this, though he is not sparing in the use of words with negative implications, such as "desert," "stillness," "silence," "nothingness."

---

[16] *The Transformation of Nature in Art*, p. 201.

Meister Eckhart and Buddhism

Perhaps when these terms are used among Western thinkers, they are understood in connection with their historical background. But as soon as these thinkers are made to plunge into a strange, unfamiliar system or atmosphere, they lose their balance and condemn it as negativistic or anarchistic or upholding escapist egoism.

According to Eckhart,

> I have read many writings both of heathen philosophers and sages, of the Old and the New Testaments, and I have earnestly and with all diligence sought the best and the highest virtue whereby man may come most closely to God and wherein he may once more become like the original image as he was in God when there was yet no distinction between God and himself before God produced creatures. And having dived into the basis of things to the best of my ability I find that it is no other than absolute detachment (*abegescheidenheit*) from everything that is created. It was in this sense when our Lord said to Martha: "One thing is needed," which is to say: He who would be untouched and pure needs just one thing, detachment.[17]

What then is the content of absolute detachment? It cannot be designated "as this or that," as Eckhart says. It is pure nothing (*bloss niht*), it is the highest point at which God can work in us as he pleases.

> Perfect detachment is without regard, without either

---

[17] Blakney, "About Disinterest," p. 82. The translator prefers "disinterest" to "detachment" for *abegescheidenheit*. I really do not know which is better. The German word seems to correspond to the Sanskrit *anabhīnivesa* or *asanga* (*mushūjaku* in Japanese and *wu chih chu* in Chinese), meaning "not attached," "not clinging to."

lowliness or loftiness to creatures; it has no mind to be below nor yet to be above; it is minded to be master of itself, loving none and hating none, having neither likeness nor unlikeness, neither this nor that, to any creature; the only thing it desires to be is to be one and the same. For to be either this or that is to want something. He who is this or that is somebody; but detachment wants altogether nothing. It leaves all things unmolested.[18]

While Buddhist emphasis is on the emptiness of all "composite things" (*skandha*) and is therefore metaphysical, Eckhart here insists on the psychological significance of "pure nothingness" so that God can take hold of the soul without any resistance on the part of the individual. But from the practical point of view the emptying of the soul making it selfless can never be thoroughly realized unless we have an ontological understanding of the nature of things, that is, the nothingness of creaturely objects. For the created have no reality; all creatures are pure nothing, for "all things were made by him [God] and without him was not anything made" (John, 1:3). Further, "If without God a creature has any being however small, then God is not the cause of all things. Besides, a creature will not be created, for creation is the receiving of being from nothing." [19] What could this mean? How could any being come from nothing or non-being? Psychology herein inevitably turns to metaphysics. We here encounter the problem of Godhead.

This problem was evidently not touched upon frequently by Eckhart, for he warns his readers re-

[18] Evans, with a little change, pp. 341–2.
[19] Blakney, pp. 298–9.

peatedly, saying: "Now listen: I am going to say something I have never said before." Then he proceeds: "When God created the heavens, the earth, and creatures, he did no work; he had nothing to do; he made no effort." He then proceeds to say something about Godhead, but he does not forget to state: "For yet again I say a thing I never said before: God and Godhead are different as earth is from heaven." Though he often fails to make a clear distinction between the two and would use "God" where really "Godhead" is meant, his attempt to make a distinction is noteworthy. With him God is still a something as long as there is any trace of movement or work or of doing something. When we come to the Godhead, we for the first time find that it is the unmoved, a nothing where there is no path (*apada*) to reach. It is absolute nothingness; therefore it is the ground of being from whence all beings come.

While I subsisted in the ground, in the bottom, in the river and fount of Godhead, no one asked me where I was going or what I was doing: there was no one to ask me. When I was flowing all creatures spake God. If I am asked, Brother Eckhart, when went ye out of your house? Then I must have been in. Even so do all creatures speak God. And why do they not speak the Godhead? Everything in the Godhead is one, and of that there is nothing to be said. God works, the Godhead does no work, there is nothing to do; in it is no activity. It never envisaged any work. God and Godhead are as different as active and inactive. On my return to God, where I am formless, my breaking through will be far nobler than my emanation. I alone take all creatures out of their sense into my mind and make them one in me. When I go back into the tasted wine in the cellar[23] may put in a question here: "How could we talk about a man's purity of heart prior to his existence? How could we also talk about seeing our own face before we were born?" Eckhart quotes St. Augustine: "There is a heavenly door for the soul into the divine nature—where somethings are reduced to nothing." [24] Evidently we have to wait for the heavenly door to open by our repeated or ceaseless knocking at it when I am "ignorant with knowing, loveless with loving, dark with light." [25] Everything comes out of this basic experience and it is only when this is comprehended that we really enter into the realm of emptiness where the Godhead keeps our discriminatory mind altogether "emptied out to nothingness." [26]

5

What is the Absolute Tao?

Before we go on to the Zen conception of the "Absolute Tao" or Godhead who sets itself up on "pure nothingness," it may be appropriate to comment on the Taoist conception of it as expounded by Lao-tzu. He was one of the early thinkers of China on philosophical subjects and the theme of the *Tao Tê Ching* ascribed to him is *Tao*.

Tao literally means "way" or "road" or "passage," and in more than one sense corresponds to the

[23] *Ibid.*, p. 216.
[24] *Ibid.*, p. 89.
[25] "*Von erkennen kennelos und von minne minnelos und von liehte vinster.*" Pfeiffer, p. 491.
[26] Blakney, p. 88.

Sanskrit *Dharma*. It is one of the key terms in the history of Chinese thought. While Taoism derives its name from this term, Confucius also uses it extensively. With the latter however it has a more moralistic than metaphysical connotation. It is Taoists who use it in the sense of "truth," "ultimate reality," "logos," etc. Lao-tzu defines it in his *Tao Tê Ching* as follows:

> The Way is like an empty vessel
> That yet may be drawn from
> Without ever needing to be filled.
> It is bottomless: the very progenitor of
>   all things in the world. . . .
> It is like a deep pool that never dries
> I do not know whose child it could be.
> It looks as if it were prior to God.[27]

There is another and more detailed characterization of Tao in Chapter XIV:

> When you look at it you cannot see it;
> It is called formless.
> When you listen to it you cannot hear it;
> It is called soundless.
> When you try to seize it you cannot hold it;
> It is called subtle.
> No one can measure these three to their
>   ultimate ends,
> Therefore they are fused to one.
>
> It is up, but it is not brightened;
> It is down, but it is not obscured.

---

[27] Translated by Arthur Waley. (From his *The Way and Its Power*, published 1934 by George Allen and Unwin Ltd. The succeeding quotations from *Tao Tê Ching* are all my rendering.) Chapter IV. God here is distinguished from Godhead as by Eckhart. The last two lines are my own version.

Meister Eckhart and Buddhism

cative. The Chinese books are best perused in large type printed from the wooden blocks.

Besides this visual appeal of the ideograms there is an element in the Chinese language which, while rare in others, especially in Indo-European languages, expresses more directly and concretely what our ordinary conceptualized words fail to communicate. For instance, read the *Tao Té Ching*, Chapter XX, in the original and compare it with any of the translations you have at hand and see that the translations invariably lack that rich, graphic, emotional flavor which we after more than two thousand five hundred years can appreciate with deep satisfaction. Arthur Waley is a great Chinese scholar and one of the best interpreters of Chinese life. His English translation of Lao-tzu is a fine piece of work in many senses, but he cannot go beyond the limitations of the language to which he is born.

6

The following story may not have historicity but it is widely circulated among Zen followers who are occasionally quite disrespectful of facts. It is worth our consideration as illustrating the way in which the Zen teachers handle the problem of "Emptiness" or "absolute nothingness" or the "still desert" lying beyond "this and that" and prior to "before and after." The story and comments are taken from a Chinese Zen textbook[29] of the Sung

[29] It is entitled *Hekigan-shu* or *Hekigan-roku* meaning "Blue Rock Collection" or "Blue Rock Records."

dynasty of the eleventh century. The text is studied very much in Japan and some of the stories are used as *kō-an* (problems given to Zen students for solution).

Bodhidharma, who is the first Zen patriarch in China, came from India in the sixth century. The Emperor Wu of the Liang dynasty invited him to his court. The Emperor Wu, a good pious Buddhist studying the various Mahāyāna *Sūtras* and practicing the Buddhist virtues of charity and humility, asked the teacher from India: "The *Sūtras* refer so much to the highest and holiest truth, but what is it, my Reverend Master?"

Bodhidharma answered, "A vast emptiness and no holiness in it."

The Emperor: "Who are you then who stand before me if there is nothing holy, nothing high in the vast emptiness of ultimate truth?"

Bodhidharma: "I do not know, your Majesty."

The Emperor failed to understand the meaning of this answer and Bodhidharma left him to find a retreat in the North.

When Bodhidharma's express purpose of coming to China was to elucidate the teaching of "vast emptiness" (*śūnyatā*), why did he answer "I do not know" to the Emperor's all-important and to-the-very-point question? It is evident, however, that Bodhidharma's answer could not have been one of an agnostic who believes in the unknowability of ultimate truth. Bodhidharma's unknowability must be altogether of a different sort. It is really what Eckhart would like to see us all have—"transformed knowledge, not ignorance which comes from lack of knowing; it is by knowing that we get to this unknowing. Then we know by divine knowing, then our ignorance is ennobled and adorned with supernatural knowledge." [30] It was this kind of unknowing which is transcendental, divine, and supernatural that he wished his imperial friend to realize.

From our ordinary relative point of view Bodhidharma may seem too abrupt and unacceptable. But the fact is that the knowledge or "I do not know" which is gained only by "sinking into oblivion and ignorance" [31] is something quite abrupt or discrete or discontinuous in the human system of knowability, for we can get it only by leaping or plunging into the silent valley of Absolute Emptiness. There is no continuity between this and the knowledge we highly value in the realm of relativity where our senses and intellect move.

The Zen teachers are all unknowing knowers or knowing unknowers. Therefore their "I do not know" does not really mean our "I do not know." We must not take their answers in the way we generally do at the level of relative knowledge. Therefore, their comments which are quoted below do not follow the line we ordinarily do. They have this unique way. Yengo (1063–1135) gives his evaluation of the *mondo* ("question and answer") which took place between Bodhidharma and the Emperor Wu of the Liang dynasty in the following words:[32]

[30] Evans, p. 13.
[31] *"Hie muoz komen in ein vergezzen und in ein nihtwizzen."* Pfeiffer, p. 14. Evans, p. 13.
[32] Yengo is given here in a modernized fashion, for the original Chinese would require a detailed interpretation.

Bodhidharma came to this country, via the southern route, seeing that there was something in Chinese mentality which responds readily to the teaching of Mahāyāna Buddhism. He was full of expectations, he wanted to lead our countrymen to the doctrine of "Mind-alone" which cannot be transmitted by letters or by means of word of mouth. The Mind could only be immediately taken hold of whereby we attain to the perception of the Buddha-nature, that is, to the realization of Buddhahood. When the Nature is attained, we shall be absolutely free from all bondage and will not be led astray because of linguistic complications. For here Reality itself is revealed in its nakedness with no kinds of veil on it. In this frame of mind Bodhidharma approached the Emperor. He also thus instructed his disciples. We see that Bodhidharma's [emptied mind] had no premeditated measures, no calculating plans. He just acted in the freest manner possible, cutting everything asunder that would obstruct his seeing directly into the Nature in its entire nakedness. Here was neither good nor evil, neither right nor wrong, neither gain nor loss. . . .

The Emperor Wu was a good student of Buddhist philosophy and wished to have the first principle elucidated by the great teacher from India. The first principle consists in the identity of being and non-being beyond which the philosophers fail to go. The Emperor wondered if this blockage could somehow be broken down by Bodhidharma. Hence his question. Bodhidharma knew that whatever answers he might give would be frustrating.

"What is Reality? What is Godhead?"

"Vast emptiness and no distinctions whatever [neither Father nor Son nor Holy Ghost]."

No philosopher however well trained in his profession could ever be expected to jump out of this trap, except Bodhidharma himself who knew perfectly well how to cut all limitations down by one blow of a sword.

Meister Eckhart and Buddhism

Most people nowadays fail to get into the ultimate signification of Bodhidharma's pronouncement and would simply cry out, "vast emptiness" as if they really experienced it. But all to no purpose! As my old master remarks, "When a man truly understands Bodhidharma, he for the first time finds himself at home quietly sitting by the fireside." Unfortunately, the Emperor Wu happened to be one of those who could not rise above the limitations of linguistics. His views failed to penetrate the screen of *meum* and *tuum* (you and me). Hence his second question: "Who are you who face me?" Bodhidharma's blunt retort, "I do not know," only helped make the august inquirer blankly stare.

Later, when he learned more about Bodhidharma and realized how stupid he was to have missed the rare opportunity of going deeper into the mystery of Reality, he was greatly upset. Hearing of Bodhidharma's death after some years he erected a memorial stele for him and inscribed on it: "Alas! I saw him, I met him, I interviewed him, and failed to recognize him. How sad! It is all past now. Alas, history is irrevocable!" He concluded his eulogy thus:

> "As long as the mind tarries on the plane of
> relativity,
> It forever remains in the dark.
> But the moment it loses itself in the Emptiness,
> It ascends the throne of Enlightenment."

After finishing the story of the Emperor Wu, Yengo the commentator puts this remark: "Tell me by the way where Bodhidharma could be located." This is expressly addressed to the readers and the commentator expects us to give him an answer. Shall we take up his challenge?

There is another commentator on this episode,

who lived some years prior to the one already referred to. This one, called Seccho (980–1052), was a great literary talent and his comments are put in a versified form full of poetic fantasies. Alluding to the Emperor Wu's attempt to send a special envoy for Bodhidharma, who after the interview crossed the Yangtzu Chiang and found a retreat somewhere in the North, the commentator goes on:

> "You [the Emperor Wu] may order all your subjects
>   to fetch him [Bodhidharma],
> But he will never show himself up again!
> We are left alone for ages to come
> Vainly thinking of the irrevocable past.
> But stop! let us not think of the past!
> The cool refreshing breeze sweeps all over the
>   earth,
> Never knowing when to suspend its work."

Seccho (the master commentator) now turns around and surveying the entire congregation (as he was reciting his versified comments), asks: "O Brethren, is not our Patriarch[33] to be discovered among us at this very moment?"

After this interruption, Seccho continues, "Yes, yes, he is here! Let him come up and wash the feet for me!"

It would have been quite an exciting event if Eckhart appeared to be present at this session which took place in the Flowery Kingdom in the first half of the eleventh century! But who can tell if Eckhart is not watching me writing this in the most modern and most mechanized city of New York?

[33] Bodhidharma.

7

A few more remarks about "Emptiness."

Relativity is an aspect[34] of Reality and not Reality itself. Relativity is possible somewhere between two or more things, for this is the way that makes one get related to another.

A similar argument applies to movement. Movement is possible in time; without the concept of time there cannot be a movement of any sort. For a movement means an object going out of itself and becoming something else which is not itself. Without the background of time this becoming is unthinkable.

Therefore, Buddhist philosophy states that all these concepts, movement and relativity, must have their field of operation, and this field is designated by Buddhist philosophers as Emptiness (*śūnyatā*).

When Buddha talks about all things being transient, impermanent, and constantly changing, and therefore teaches that there is nothing in this world which is absolutely dependable and worth clinging to as the ultimate seat of security, he means that we must look somewhere else for things permanent (*jō*), bliss-imparting (*raku*), autonomous (*ga*), and absolutely free from defilements (*jō*). According to the *Nirvāna Sūtra* (of the Mahāyāna school), these four (*jō-raku-ga-jō*) are the qualities of Nirvana, and Nirvana is attained when we have knowledge, when the mind is freed from thirst (*tanhā*), cravings (*āsava*), and conditionality (*sankhāra*). While Nir-

---
[34] *Sō* in Japanese, *hsiang* in Chinese, *lakṣaṇa* in Sanskrit.

vana is often thought to be a negativistic idea the Mahāyāna followers have quite a different interpretation. For they include autonomy (*ga, ātman*) as one of its qualities (*guna*), and autonomy is free will, something dynamic. Nirvana is another name for the Emptiness.

The term "emptiness" is apt to be misunderstood for various reasons. The hare or rabbit has no horns, the turtle has no hair growing on its back. This is one form of emptiness. The Buddhist *śūnyatā* does not mean absence.

A fire has been burning until now and there is no more of it. This is another kind of emptiness. Buddhist *śūnyatā* does not mean extinction.

The wall screens the room: on this side there is a table, and on the other side there is nothing, space is unoccupied. Buddhist *śūnyatā* does not mean vacancy.

Absence, extinction, and unoccupancy—these are not the Buddhist conception of emptiness. Buddhists' Emptiness is not on the plane of relativity. It is Absolute Emptiness transcending all forms of mutual relationship, of subject and object, birth and death, God and the world, something and nothing, yes and no, affirmation and negation. In Buddhist Emptiness there is no time, no space, no becoming, no-thing-ness; it is what makes all these things possible; it is a zero full of infinite possibilities, it is a void of inexhaustible contents.

Pure experience is the mind seeing itself as reflected in itself, it is an act of self-identification, a state of suchness. This is possible only when the mind is *śūnyatā* itself, that is, when the mind is

Meister Eckhart and Buddhism

devoid of all its possible contents except itself. But to say "except itself" is apt to be misunderstood again. For it may be questioned, what is this "itself"? We may have to answer in the same way as St. Augustine did: "When you ask, I do not know; but when you do not, I know."

The following dialogue which took place between two Zen masters of the T'ang dynasty will help show us what methodology was adopted by Zen for communicating the idea of "itself."

One master called Isan (771–853) was working with his disciples in the garden, picking tea leaves. He said to one of the disciples in the garden called Kyōzan, who also was a master: "We have been picking the tea leaves all day; I hear your voice only and do not see your form. Show me your primeval form." Kyōzan shook the tea bushes. Isan said, "You just got the action, but not the body." Kyōzan then said, "What would be your answer?" Isan remained quiet for a while. Thereupon Kyōzan said, "You have got the body, but not the action." Isan's conclusion was, "I save you from twenty blows of my stick."

As far as Zen philosophy is concerned this may be all right, as these two masters know what each is seeking to reveal. But the business of philosophers of our modern epoch is to recognize or to probe the background of experience on which these Zen masters stand and try to elucidate it to the best of their capacity. The masters are not simply engaged in mystifying the bystanders.

To say "empty" is already denying itself. But you cannot remain silent. How to communicate the silence without going out of it is the crux. It is for this reason that Zen avoids as much as possible resorting to linguistics and strives to make us go underneath words, as it were to dig out what is there. Eckhart is doing this all the time in his sermons. He picks out some innocent words from the Bible and lets them disclose an "inner act" which he experiences in his unconscious consciousness. His thought is not at all in the words. He turns them into instruments for his own purposes. In a similar way the Zen master makes use of anything about himself including his own person, trees, stones, sticks, etc. He may then shout, beat, or kick. The main thing is to discover what is behind all these actions. In order to demonstrate that Reality is "Emptiness," the Zen master may stand still with his hands folded over his chest. When he is asked a further question, he may shake the tea plant or walk away without a word, or give you a blow of a stick.

Sometimes the master is more poetic and compares the mind of "emptiness" to the moon, calling it the mind-moon or the moon of suchness. An ancient master of Zen philosophy sings of this moon:

> The mind-moon is solitary and perfect:
> The light swallows the ten-thousand things.
> It is not that the light illuminates objects,
> Nor are objects in existence.
> Both light and objects are gone,
> And what is it that remains?

The master leaves the question unanswered. When it is answered the moon will no longer be there.

Meister Eckhart and Buddhism

Reality is differentiated and Emptiness vanishes into an emptiness. We ought not to lose sight of the original moon, primeval mind-moon, and the master wants us to go back to this, for it is where we have started first. Emptiness is not a vacancy, it holds in it infinite rays of light and swallows all the multiplicities there are in this world.

Buddhist philosophy is the philosophy of "Emptiness," it is the philosophy of self-identity. Self-identity is to be distinguished from mere identity. In an identity we have two objects for identification; in self-identity there is just one object or subject, one only, and this one identifies itself by going out of itself. Self-identity thus involves a movement. And we see that self-identity is the mind going out of itself in order to see itself reflected in itself. Self-identity is the logic of pure experience or of "Emptiness." In self-identity there are no contradictions whatever. Buddhists call this suchness.

I once talked with a group of lovers of the arts on the Buddhist teaching of "Emptiness" and Suchness, trying to show how the teaching is related to the arts. The following is part of my talk.

To speak the truth, I am not qualified to say anything at all about the arts, because I have no artistic instincts, no artistic education, and have not had many opportunities to appreciate good works of art. All that I can say is more or less conceptual.

Take the case of painting. I often hear Chinese or Japanese art critics declare that Oriental art consists in depicting spirit and not form. For they say that when the spirit is understood the form creates itself; the main thing is to get into the spirit of an object which the painter chooses for his subject. The West, on the other hand, emphasizes form, endeavors to reach the spirit by means of form. The East is just the opposite: the spirit is all in all. And it thinks that when the artist grasps the spirit, his work reveals something more than colors and lines can convey. A real artist is a creator and not a copyist. He has visited God's workshop and has learned the secrets of creation—creating something out of nothing.

With such a painter every stroke of his brush is the work of creation, and it cannot be retraced because it never permits a repetition. God cannot cancel his fiat; it is final, irrevocable, it is an ultimatum. The painter cannot reproduce his own work. When even a single stroke of his brush is absolute, how can the whole structure or composition be reproduced, since this is the synthesis of all his strokes, every one of which has been directed towards the whole?

In the same way every minute of human life as long as it is an expression of its inner self is original, divine, creative, and cannot be retrieved. Each individual life thus is a great work of art. Whether or not one makes it a fine inimitable masterpiece depends upon one's consciousness of the working of *śūnyatā* within oneself.

How does the painter get into the spirit of the plant, for instance, if he wants to paint a hibiscus as Mokkei (Mu-chi) of the thirteenth century did in his famous picture, which is now preserved as a national treasure at Daitokuji temple in Kyoto? The secret is to become the plant itself. But how

Meister Eckhart and Buddhism

can a human being turn himself into a plant? Inasmuch as he aspires to paint a plant or an animal, there must be in him something which corresponds to it in one way or another. If so, he ought to be able to become the object he desires to paint.

The discipline consists in studying the plant inwardly with his mind thoroughly purified of its subjective, self-centered contents. This means to keep the mind in unison with the "Emptiness" or Suchness, whereby one who stands against the object ceases to be the one outside that object but transforms himself into the object itself. This identification enables the painter to feel the pulsation of one and the same life animating both him and the object. This is what is meant when it is said that the subject is lost in the object, and that when the painter begins his work it is not he but the object itself that is working and it is then that his brush, as well as his arm and his fingers, become obedient servants to the spirit of the objects. The object makes its own picture. The spirit sees itself as reflected in itself. This is also a case of self-identity.

It is said that Henri Matisse looked at an object which he intended to paint for weeks, even for months, until its spirit began to move him, to urge him, even to threaten him, to give it an expression.

A writer on modern art, I am told, says that the artist's idea of a straight line is different from that of the mathematician, for the former conceives a straight line as fusing with a curve. I do not know whether this quotation is quite correct, but the remark is most illuminating. For a straight line that remains always straight is a dead line and the curve that cannot be anything else is another dead line. If they are at all living lines, and this ought to be the case with every artistic production, a straight line is curved and a curve is straight; besides there ought to be what is known as "dimensional tension" in every line. Every living line is not just on one plane, it is suffused with blood, it is tridimensional.

I am also told that color with the artist is not just red or blue, it is more than perceptual, it is charged with emotion. This means that color is a living thing with the artist. When he sees red it works out its own world; the artist bestows a heart on the color. The red does not stop just at being one of the seven colors as decomposed through the prism. As a living thing it calls out all other colors and combines them in accordance with its inner promptings. Red with the artist is not a mere physical or psychological event, it is endowed with a spirit.

These views are remarkably Oriental. There is another striking statement made by a Western artist. According to him, when he is thoroughly absorbed in a visual perception of any kind, he feels within himself certain possibilities out of the visual representation which urge him to give them an expression. The artist's life is that of the creator. God did not make the world just for the sake of making something. He had a certain inner urge, he wanted to see himself reflected in his creation. That is what is meant when the Bible speaks about God's making man after his own likeness. It is not man alone that is God's image, the whole world is his image, even the meanest flea as Eckhart would say shares God's is-ness in its is-ness. And because of this is-ness the whole world moves on. So with the artist. It is due to his is-ness being imbued into his works that they are alive with his spirit. The artist himself may not be conscious of all this proceeding, but Zen knows and is also prepared to impart the knowledge to those who would approach it in the proper spirit. The is-ness of a thing is not just being so, but it contains in it infinite possibilities which Buddhists call *tê* in Chinese, *toku* in Japanese, and *guna* in Sanskrit. This is where lies "the mystery of being," which is "the inexhaustibility of the Emptiness."

The following story of Rakan Osho (Lohan Hoshang), of Shōshu, China, who lived in the ninth century, is given here to illustrate how Zen transforms one's view of life and makes one truly see into the is-ness of things. The verse relates his own experience.

It was in the seventh year of Hsien-t'ung [867 A.D.]
  that I for the first time took up the study of the Tao [that is, Zen].

Whenever I went I met words and did not understand them.
A lump of doubt inside the mind was like a willow-basket.
For three years, residing in the woods by the stream, I was altogether unhappy.
When unexpectedly I happened to meet the Dharmarāja [Zen master] sitting on the rug,
I advanced towards him earnestly asking him to dissolve my doubt.
The master rose from the rug on which he sat deeply absorbed in meditation;

He then baring his arm gave me a blow with his fist on my chest.
This all of a sudden exploded my lump of doubt completely to pieces.
Raising my head I for the first time perceived that the sun was circular.
Since then I have been the happiest man in the world, with no fears, no worries;
Day in day out, I pass my time in a most lively way.
Only I notice my inside filled with a sense of fullness and satisfaction;
I do not go out any longer, hither and thither, with my begging bowl for food.[35]

What is of the most significant interest in his verse-story of Rakan Osho's experience is that "he for the first time perceived that the sun was round." Everybody knows and sees the sun and the Osho also must have seen it all his life. Why then does he specifically refer to it as circular as if he saw it really for the first time? We all think we are living, we really eat, sleep, walk, talk. But are we really? If we were, we would never be talking about "dread," "insecurity," "fear," "frustration," "courage to be," "looking into the vacant," "facing death."

---

[35] *The Transmission of the Lamp* (*Dentoroku*), fas. XI.

chapter 2

The Basis of Buddhist Philosophy

I

Buddhist philosophy is based on the experience Buddha had about twenty-five centuries ago. To understand, therefore, what Buddhist philosophy is, it is necessary to know what that experience was which Buddha had after six years' hard thinking and ascetic austerities and exercises in meditation.

We generally think that philosophy is a matter of pure intellect, and, therefore, that the best philosophy comes out of a mind most richly endowed with intellectual acumen and dialectical subtleties. But this is not the case. It is true that those who are poorly equipped with intellectual powers cannot be good philosophers. Intellect, however, is not the whole thing. There must be a deep power of imagination, there must be a strong, inflexible will-power, there must be a keen insight into the nature of man, and finally there must be an actual seeing of the truth as synthesized in the whole being of the man himself.

I wish to emphasize this idea of "seeing." It is not enough to "know" as the term is ordinarily understood. Knowledge unless it is accompanied by a personal experience is superficial and no kind of philosophy can be built upon such a shaky foundation. There are, however, I suppose many systems of thought not backed by real experiences, but such are never inspiring. They may be fine to look at but their power to move the readers is nil. Whatever knowledge the philosopher may have, it must come out of his experience, and this experience is seeing. Buddha has always emphasized this. He couples knowing (*ñāṇa, jñāna*) with seeing (*passa, paśya*), for without seeing, knowing has no depths, cannot understand the realities of life. Therefore, the first item of the Eightfold Noble Path is *sammādassana*, right seeing, and *sammāsankappa*, right knowing, comes next. Seeing is experiencing, seeing things in their state of suchness (*tathatā*) or is-ness. Buddha's whole philosophy comes from this "seeing," this experiencing.

The experience which forms the basis of Buddhist philosophy is called "enlightenment-experience," for it is this experience of enlightenment which Buddha had after six years of hard thinking and profound reflection, and everything he taught afterward is the unfolding of this inner perception he then had.

What then was this enlightenment-experience?

2

Roughly speaking, we can say that there are two ways of approaching this question: What is the enlightenment-experience Buddha had? One is objective and the other subjective. The objective approach is to find out the first rationalized statements ascribed to Buddha after the experience and understood as forming the basis of his teaching. That is, what did he first teach? What was the main thesis he continued to preach throughout his life? This

The Basis of Buddhist Philosophy

will be to discover what characteristically constitutes the Buddhist teaching as distinguished from that of the rest of the Indian thinkers. The second approach, called subjective, is to examine Buddha's utterances reflecting his immediate feelings after the experience of enlightenment. The first approach is metaphysical whereas the second is psychological or existential. Let us start with the first.

What is universally recognized as Buddhist thought regardless of its varieties of interpretation is the doctrine of *anattā* or *anātman*, that is, the doctrine of non-ego. Its argument begins with the idea: (1) that all things are transient as they are composites (*skandha* or *khandha*) and go on disintegrating all the time, that there is nothing permanent; and (2) that there is therefore nothing worth clinging to in this world where every one of us is made to undergo all kinds of sorrow and suffering. How do we escape from them? Or, how do we conquer them? For we cannot go on like this. We must somehow find the way out of this torture. It was this feeling of fear and insecurity individually and collectively that made Buddha leave his home and wander about for six long years seeking for a way out not only for himself but for the whole world. He finally discovered it by hitting upon the idea of non-ego (*anattā*). The formula runs thus:[1]

All composite things (*sankhāra*) are impermanent. When a man by wisdom (*paññā*) realizes [this,], he heeds not [this world of] sorrow; this is the path to purity.

[1] *The Dhammapada*, translated by S. Radhakrishnan (Oxford University Press, 1951), verses 277–9, pp. 146–7. I do not, however, always follow him in my quotations in this book.

All composite things are sorrowful. When a man by wisdom realizes [this], he heeds not [this world of] sorrow; this is the path to purity.

All things (*dhammā*) are egoless. When a man by wisdom realizes [this], he heeds not [this world of] sorrow; this is the path to purity.

The one thing I wish to call to the readers' attention is the term "wisdom," *paññā*, or *prajñā* in Sanskrit. This is a very important term throughout Buddhist philosophy. There is no English equivalent for it. "Transcendental wisdom" is too heavy, besides it does not exactly hit the mark. But temporarily let "wisdom" do. We know that seeing is very much emphasized in Buddhism, but we must not fail also to notice that seeing is not just an ordinary seeing by means of relative knowledge; it is the seeing by means of a *prajñā*-eye which is a special kind of intuition enabling us to penetrate right into the bedrock of Reality itself. I have elsewhere[2] given a somewhat detailed account of *prajñā* and its role in Buddhist teachings, especially in Zen Buddhism.

The doctrine of non-ego not only repudiates the idea of an ego-substance but points out the illusiveness of the ego-idea itself. As long as we are in this world of particular existences we cannot avoid cherishing the idea of an individual ego. But this by no means warrants the substantiality of the ego. Modern psychology has in fact done away with an ego-entity. It is simply a workable hypothesis by which we carry on our practical business. The prob-

[2] *Studies in Zen* (London: Rider and Company, 1955), 85–128.

The Basis of Buddhist Philosophy

lem of the ego must be carried on to the field of metaphysics. To really understand what Buddha meant by saying that there is no *ātman*, we must leave psychology behind. Because it is not enough just to state that there is no *ātman* if we wish really to reach the end of sorrow and to be thus at peace with ourselves and with the world at large. We must have something positive in order to see ourselves safely in the harbor and securely anchored. Mere psychology cannot give us this. We must go out to a broader field of Reality where *prajñā*-intuition comes into play.

As long as we wander in the domain of the senses and intellect, the idea of an individual ego besets us, and makes us eternally pursue the shadow of the ego. But the ego is something always eluding our grasp; when we think we have caught it, it is found to be no more than a slough left by the snake while the real ego is somewhere else. The human ego-snake is covered with an infinity of sloughs, the catcher will before long find himself all exhausted. The ego must be caught not from outside but from within. This is the work of *prajñā*. The wonder *prajñā* performs is to catch the actor in the midst of his action, he is not made to stop acting in order to be seen as actor. The actor is the acting, and the acting is the actor, and out of this unification or identification *prajñā* is awakened. The ego does not go out of himself in order to see himself. He stays within himself and sees himself as reflected in himself. But as soon as a split takes place between the ego as actor and the ego as seer or spectator, *prajñā* is dichotomized, and all is lost.

Eckhart expresses the same experience in terms of Christian theology. He talks about Father, Son, Holy Ghost, and love. They sound unfamiliar to Buddhist ears but when they are read with a certain insight we will find that "the love with which he [God] loves himself" is the same as the *prajñā*-intuition that sees into the ego itself. Eckhart tells us: "In giving us his love God has given us his Holy Ghost so that we can love him with the love wherewith he loves himself. We love God with his own love; awareness of it deifies us." [3] The Father loving the Son and the Son loving the Father—this mutual love, that is, love loving itself is, in Zen terminology, one mirror reflecting another with no shadow between them. Eckhart calls this "the play going on in the Father-nature. Play and audience are the same." He continues:

This play was played eternally before all natures. As it is written in the Book of Wisdom, "Prior to creatures, in the eternal now, I have played before the Father in an eternal stillness." The Son has eternally been playing before the Father as the Father has before his Son. The playing of the twain is the Holy Ghost in whom they both disport themselves and he disports himself in both. Sport and players are the same. Their nature proceeding in itself. "God is a fountain flowing into itself," as St. Dionysius says.

Prajñā-intuition comes out of itself and returns to itself. The self or ego that has been constantly eluding our rationalized scrutiny is at last caught when it comes under *prajñā*-intuition which is no other than the self.

[3] Evans, pp. 147 ff.

The Basis of Buddhist Philosophy 45

Buddhists generally talk about the egolessness (*anattā* or *anātmya*) of all things, but they forget that the egolessness of things cannot really be understood until they are seen with the eye of *prajñā*-intuition. The psychological annihilation of an ego-substance is not enough, for this still leaves the light of *prajñā*-eye under a coverage. Eckhart says, "God is a light shining itself in silent stillness." (Evans, p. 146.) As long as our intellectually analytic eye is hotly pursuing the shadow of Reality by dichotomizing it, there will be no silent stillness of absolute identity where *prajñā* sees itself reflected in itself. Eckhart is in accord with the Buddhist experience when he proceeds: "The Word of the Father is none other than his understanding of himself. The understanding of the Father understands that he understands, and that his understanding understands is the same as that he is who is understanding. That is, the light from the light." (*Ibid.*, p. 146.)

The psychological analysis that cannot go further or deeper than the egolessness of the psychological ego fails to see into the egolessness of all things (*dharma*), which appears to the eye of *prajñā*-intuition not as something sheerly of privative value but as something filled with infinite possibilities. It is only when the *prajñā*-eye surveys the nature of all things (*sarvadharma* or *sabbe dhamma*), that their egolessness displays positive constructive energies by first dispelling the clouds of Māyā, by demolishing every structure of illusion, and thus finally by creating a world of altogether new values based on *prajñā* (wisdom) and *karuṇā* (love). The enlightenment-experience therefore means going beyond the world of psychology, the opening of the *prajñā*-eye, and seeing into the realm of Ultimate Reality, and landing on the other shore of the stream of *samsāra*, where all things are viewed in their state of suchness, in the way of purity. This is when a man finds his mind freed from everything (*sabbattha vimuttamānasa*),[4] not confounded by the notions of birth-and-death, of constant change, of before, behind, and middle. He is the "conqueror" to whom *The Dhammapada* (179) gives this qualification:

He whose conquest nobody can conquer again,
Into whose conquest nobody in this world can enter—
By what track can you trace him,
The awakened, of infinite range, the trackless?

Such an awakened one is an absolute conqueror and nobody can follow his tracks as he leaves none. If he leaves some, this will be turned into the means whereby he can be defeated. The realm where he lives has no limiting boundaries, it is like a circle whose circumference is infinite, therefore with no center to which a path can lead. This is the one Zen describes as a man of *anābhogacaryā* ("an effortless, purposeless, useless man").[5] This corresponds to Eckhart's man of freedom who is defined as "one who clings to nothing and to whom nothing clings" (Evans, p. 146). While these statements are apt to suggest the doctrine of doing-nothing-ness we must remember that Buddhists are great adherents

[4] *The Dhammapada*, verse 348, p. 167.
[5] *Studies in the Lankāvatāra Sūtra*, pp. 223 ff.

The Basis of Buddhist Philosophy 47

of what is known as the teaching of *karuṇā* and *praṇidhāna*, to which the reader is referred below.

3

When "the egolessness of all things seen with *prajñā*," [6] which makes us transcend sorrows and sufferings and leads to "the path of purity," is understood in the sense herein elucidated, we find the way to the understanding of the lines known as "hymn of victory."

The hymn is traditionally ascribed to Buddha who uttered it at the time of his enlightenment. It expresses more of the subjective aspect of his experience which facilitates our examination of the content of the enlightenment. While the egolessness of things is Buddha's metaphysical interpretation of the experience as he reflected upon it, the hymn of victory echoes his immediate reaction, and we are able to have a glimpse into the inner aspect of Buddha's mind more directly than through the conceptualization which came later. We can now proceed to what I have called the second approach. The hymn runs as follows:

Looking for the maker of this tabernacle
I ran to no avail
Through a round of many births;
And wearisome is birth again and again.
But now, maker of the tabernacle, thou hast been seen;
Thou shalt not rear this tabernacle again.
All thy rafters are broken,

[6] "*Sabbe dhammā anattā' ti yadā paññāya passati.*"

Thy ridge-pole is shattered;
The mind approaching the Eternal,
Has attained to the extinction of all desires.[7]

This is Irving Babbitt's translation, the lines of which were rearranged according to the original Pali. Incidentally, I wish to remark that there is one point in it which is unsatisfactory from my point of view. This is the phrase, "the mind approaching the Eternal." The original is *"visankhāragatam cittaṃ."* This means "the mind released from its binding conditions." "Approaching the Eternal" is the translator's own idea read into the line. Henry Warren, author of *Buddhism in Translations,* translates it "this mind has demolition reached," which points to nihilism or negativism, while Babbitt's translation has something of positive assertion. The difference so conspicuous in these two translations shows that each interprets the meaning according to his own philosophy. In this respect my understanding, which is given below, also reflects my own thought as regards the significance of Buddhist teaching generally.

The most essential thing here is the experience that Buddha had of being released from the bondage in which he had been kept so long. The utmost consciousness that filled his mind at the time of enlightenment was that he was no longer the slave to what he calls "the maker of the tabernacle," or "the builder of this house," that is, *gahakāraka*. He

[7] *The Dhammapada*, pp. 153–4. (Published by Oxford University Press, 1936.)

The Basis of Buddhist Philosophy 49

now feels himself to be a free agent, master of himself, not subject to anything external; he no longer submits himself to dictation from whatever source it may come. The *gahakāraka* is discovered, the one who was thought to be behind all his mental and physical activities, and who, as long as he, that is, Buddha, was ignorant, made him a slave to this autocrat, and employed Buddha—in fact anybody who is ignorant of the *gahakāraka*—to achieve the latter's egocentric impulses, desires, cravings. Buddha was an abject creature utterly under the control of this tyrant, and it was this sense of absolute helplessness that made Buddha most miserable, unhappy, and given over to all kinds of fears, dejection, and moroseness. But Buddha now discovers who this *gahakāraka* is; not only does he know him, but he has actually seen him face to face, taken hold of him at work. The monster, the housebuilder, the constructor of the prison-house, being known, being seen, being caught, ceases at last to weave his entrapping network around Buddha. This means what the phrase *"visankhāragatam cittam"* means, the mind freed from the bondage of its conditioning aggregates (*sankhāra*).

We must however remember that the *gahakāraka* is not dead, he is still alive, for he will be living as long as this physical existence continues. Only he has ceased to be my master; on the contrary, I am his master, I can use him as I wish, he is ready now to obey my command. "Being free from the tyranny of its binding conditions" does not mean that the conditions no longer exist. As long as we are relative existences we are to that extent conditioned, but the knowledge that we are so conditioned transcends the conditions and thus we are above them. The sense of freedom arises from this, and freedom never means lawlessness, wantonness, or libertinism. Those who understand freedom in this latter sense and act accordingly are making themselves slaves to their egotistic passions. They are no longer masters of themselves but most despicable slaves of the *gahakāraka*.

The seeing of the *gahakāraka* therefore does not mean the "seeing of the last of all desire," nor is it "the extinction of all desires." It only means that all the desires and passions we are in possession of, as human beings, are now under the control of one who has caught the *gahakāraka* working out his own limited understanding of freedom. The enlightenment-experience does not annihilate anything; it sees into the working of the *gahakāraka* from a higher point of understanding, which is to say, by means of *prajñā*, and arranges it where it properly belongs. By enlightenment Buddha sees all things in their proper order, as they should be, which means that Buddha's insight has reached the deepest depths of Reality.

As I have said before, the seeing plays the most important role in Buddhist epistemology, for seeing is at the basis of knowing. Knowing is impossible without seeing; all knowledge has its origin in seeing. Knowing and seeing are thus found generally united in Buddha's teaching. Buddhist philosophy therefore ultimately points to seeing reality as it is.

The Basis of Buddhist Philosophy

Seeing is experiencing enlightenment. The *Dharma*[8] is predicated as *ehipassika*, the *Dharma* is something "you come and see." It is for this reason that *sammādassana* (*sammādiṭṭhi* in Sanskrit) is placed at the beginning of the Eightfold Noble Path.

What is the *gahakāraka*?

The *gahakāraka* detected is our relative, empirical ego, and the mind freed from its binding conditions (*sankhāra*) is the absolute ego, *Atman*, as it is elucidated in the *Nirvāna Sūtra*. The denial of *Atman* as maintained by earlier Buddhists refers to *Atman* as the relative ego and not to the absolute ego, the ego after enlightenment-experience.

Enlightenment consists in seeing into the meaning of life as the relative ego and not as the absolute ego, the ego after enlightenment-experience.

Enlightenment consists in seeing into the meaning of life as the interplay of the relative ego with the absolute ego. In other words, enlightenment is seeing the absolute ego as reflected in the relative ego and acting through it.

Or we may express the idea in this way: the absolute ego creates the relative ego in order to see itself reflected in it, that is, in the relative ego. The absolute ego, as long as it remains absolute, has no means whereby to assert itself, to manifest itself, to work out all its possibilities. It requires a *gahakāraka* to execute its biddings. While the *gahakāraka* is not to build his tabernacle according to his own

---

[8] *Dhamma* in Pali. It has a multiple meaning and is difficult to render it uniformly. Here it stands for Truth, Reality, Norm.

design, he is an efficient agent to actualize whatever lies quiescently in the *Atman* in the sense of the *Nirvāna Sūtra*.

4

The question now is: Why does the absolute *Atman* want to see itself reflected in the empirical *Atman*? Why does it want to work out its infinite possibilities through the empirical *Atman*? Why does it not remain content with itself instead of going out to a world of multitudes, thereby risking itself to come under the domination of *sankhāra*? This is making itself, as it were, a willing slave of the *gahakāraka*.

This is a great mystery which cannot be solved on the plane of intellection. The intellect raises the question, but fails to give it a satisfactory solution. This is in the nature of the intellect. The function of the intellect consists in leading the mind to a higher field of consciousness by proposing all sorts of questions which are beyond itself. The mystery is solved by living it, by seeing into its working, by actually experiencing the significance of life, or by tasting the value of living.[9]

Tasting, seeing, experiencing, living—all these demonstrate that there is something common to enlightenment-experience and our sense-experience; the one takes place in our innermost being, the other on the periphery of our consciousness. Personal experience is thus seen to be the foundation of

---

[9] "O taste and see that the Lord is good; blessed is the man that trusteth in him." (Psalms, 34:8.)

The Basis of Buddhist Philosophy

Buddhist philosophy. In this sense Buddhism is radical empiricism or experientialism, whatever dialectic later developed to probe the meaning of enlightenment-experience.

Buddhist philosophy has long been wrongly regarded as nihilistic and not offering anything constructive. But those who really try to understand it and are not superficially led to misconstrue such terms as demolition, annihilation, extinction, breaking up, cessation, or quiescence, or without thirst, cutting off lust and hatred, will readily see that Buddha never taught a religion of "eternal death."

"Eternal death," which is sometimes regarded as the outcome of the Buddhist idea of egolessness, is a strange notion making no sense whatever. "Death" can mean something only when it is contrasted to birth, for it is a relative term. Eternal death is squaring a circle. Death never takes place unless there is a birth. Where there is birth there is death; where there is death there is birth; birth and death go together. We can never have just one of them, leaving out the other. Where there is eternal death there must be continuous birth. Where eternal death is maintained there must be a never-ceasing birth. Those who talk about total annihilation or extinction as if such things were possible are those who have never faced facts of experience.

Life is a never-ending concatenation of births and deaths. What Buddhist philosophy teaches is to see into the meaning of life as it flows on. When Buddhists declare that all things are impermanent, subject to conditions, and that there is nothing in this world of *samsāra* (birth-and-death) which can give us the hope for absolute security, they mean that as long as we take this world of transiency as something worth clinging to we are sure to lead a life of frustration. To transcend this negativistic attitude toward life we must make use of *prajñā* which is the way of purity. We must see things with the eye of *prajñā*, not to deny them as rubbish but to understand them from an aspect closed to ordinary observers. The latter see nothing but the impermanence or transiency or changeability of things and are unable to see eternity itself that goes along with time-serialism which can never be demolished. The demolition is on our side and not on the side of time. Buddha's enlightenment-experience clearly points to this. The ridgepole smashed and the rafters torn down all belong to time-serialism and not to eternity which suffers no kind of demolition. To imagine that when serialism is transcended eternity goes out of sight as if it were something relatively coexistent with time is altogether an erroneous way to interpret Buddha's utterance. It really requires the *prajñā-eye* to see into the "*sankhāra*-freed mind," which is in fact no other than Eckhart's eye: "The eye wherein I see God is the same eye wherein God sees me: my eye and God's eye are one eye, one vision, one knowing, one love." Time is eternity and eternity is time. In other words, zero is infinity and infinity is zero. The way of purity opens when the eye sees inwardly as well as outwardly—and this simultaneously. The *prajñā* seeing is one act, one glimpse, one *cittakṣāna* which is no *cittakṣāna*. Unless this truth is seen with *prajñā*-intuition, the "hymn of victory" will never yield its full meaning.

The Basis of Buddhist Philosophy

Those who read it otherwise cannot go beyond negativism or nihilism.

The following from Eckhart will shed much light:

> Renewal befalls all creatures under God; but for God there is no renewal, only all eternity. What is eternity?—It is characteristic of eternity that in it youth and being are the same, for eternity would not be eternal could it newly become and were not always.[10]

"Renewal" means "becoming" which is "transiency." What is eternal never knows "renewal," never grows old, remains forever "youthful," and transcends "demolition" or "annihilation" of all kinds. Enlightenment is to know what this "eternity" is, and this knowing consists in "knowing eternity-wise his [God's] is-ness free from becoming, and his nameless nothingness" [11] Eckhart is quite definite in giving us what kind of God he has in mind in this matter of knowing and not knowing:

> Know'st thou of him anything? He is no such thing, and in that thou dost know of him anything at all thou art in ignorance, and ignorance leads to the condition of the brute; for in creatures what is ignorant is brutish. If thou wouldst not be brutish then, know nothing of the unuttered God.—"What then shall I do?"—Thou shalt lose thy thy-ness and dissolve in his his-ness; thy thine shall be his mine, so utterly one mine that thou in him shalt know eternalwise his is-ness, free from becoming: his nameless nothingness.[12]

Eckhart's God of nameless nothingness is in Bud-

---
[10] Evans, p. 246.
[11] Pfeiffer, p. 319. *"Du mit ime verstandest ewicliche sine ungewordene istikeit under sine ungenanten nihtheit."*
[12] Evans, p. 246.

dhist terms no other than the egolessness of all things, the *sankhāra*-free mind, and the cessation of all cravings.

5

In this connection I think it is opportune to say a few words about the negative statements liberally used in Buddhist and other texts dealing with problems of ultimate reality. I may also touch a little on the frequency of paradoxical propositions used to express a certain experience popularly known as mystic.

Considering all in all, there are two sources of knowledge, or two kinds of experience, or "two births of man" as Eckhart has it, or two forms of truth (*satyā*) according to the upholders of the "Emptiness" doctrine (*śūnyavāda*). Unless this is recognized we can never solve the problem of logical contradiction which when expressed in words characterizes all religious experiences. The contradiction so puzzling to the ordinary way of thinking comes from the fact that we have to use language to communicate our inner experience which in its very nature transcends linguistics. But as we have so far no means of communication except the one resorted to by followers of Zen Buddhism, the conflicts go on between rationalists and so-called mystics. Language developed first for the use of the first kind of knowledge which was highly utilitarian, and for this reason it came to assert itself over all human affairs and experiences. Its overwhelming authority is such that we have almost come to accept everything language commands. Our thoughts have now to be molded according to its dictates, our acts are to be regulated by the rules it came to formulate for its own effective operation. This is not all. What is worse is that language has now come even to suppress the truth of new experiences, and that when they actually take place, it will condemn them as "illogical" or "unthinkable" and therefore as false, and finally that as such it will try to put aside anything new as of no human value.

The Śūnyatā school distinguishes two forms of truth (*satyā*): (1) *samvṛitti* of the relative world and (2) *paramārtha* of the transcendental realm of *prajñā*-intuition. When Buddha speaks of his enlightenment in the *Saddharmapuṇḍarīka Sūtra* ("Lotus Gospel"), he describes his experience as something which cannot be comprehended by any of his followers because their understanding can never rise up to the level of Buddha's. It is another Buddha who understands a Buddha, Buddhas have their own world into which no beings of ordinary caliber of mentality can have a glimpse. Language belongs to this world of relativity, and when Buddha tries to express himself by this means his hearers are naturally barred from entering his inner life. While in the *Laṅkāvatāra Sūtra* we are told of many other Buddha-countries where Buddha-activities are carried on by means other than mere language, for instance, by moving hands or legs, by smiling, by coughing, by sneezing, etc. Evidently Buddhas can understand one another by whatever means they may employ in conveying their inner acts, because they all know what they are through their experience. But where there are no such corresponding experiences, no amount of technique one may resort to will be possible to awaken them in others.

In Aśvaghoṣa's *Awakening of Faith* reference is made to two aspects of *Tathatā* ("Suchness") one of which is altogether beyond speaking or writing, because it does not fall into the categories of communicability. Language here has no use whatever. But Aśvaghoṣa continues: if we did not appeal to language there is no way to make others acquainted with the absolute; therefore language is resorted to in order to serve as a wedge in getting out the one already in use; it is like a poisonous medicine to counteract another. It is a most dangerous weapon and its user has to be cautioned in every way not to hurt himself. The *Lankāvatāra* is decisive in this respect:

> ... word-discrimination cannot express the highest reality, for external objects with their multitudinous individual marks are non-existent, and only appear before us as something revealed out of Mind itself. Therefore, Mahāmati, you must try to keep yourself away from the various forms of word-discrimination.[13]

Word-discrimination belongs to the *samvṛitti*, to things of the relative world, and is not meant for communicating anything that goes beyond this world of numbers and multiplicities. For here language ceases to be supreme and must realize that it has its limitations. Two of the three kinds of knowledge distinguished by Eckhart are of the

---

[13] *Lankāvatāra Sūtra*, translated by D. T. Suzuki (London: George Routledge and Sons, Ltd., 1932), p. 77.

The Basis of Buddhist Philosophy

Saṃvṛitti, whereas the third corresponds to the *paramārtha*. To quote Eckhart:

> These three things stand for three kinds of knowledge. The first is sensible. The eye sees from afar what is outside it. The second is rational and is a great deal higher. The third corresponds to an exalted power of the soul, a power so high and noble it is able to see God face to face in his own self. This power has naught in common with naught, it knows no yesterday or day before, no morrow or day after (for in eternity there is no yesterday or morrow): therein it is the present now; the happenings of the thousand years ago, a thousand years to come, are there in the present and the antipodes the same as here.[14]

The first two kinds apply to the world of senses and the intellect where language has its utmost usefulness. But when we try to use it in the realm where "the exalted power of the soul" has its sway it miserably fails to convey the activities going on there to those whose "power" has never been "heightened" or enhanced to the level indicated by Eckhart. But as we are forced to make use of language inasmuch as we are creatures of the sense-intellect, we contradict ourselves, as we see in Eckhart's statements just quoted. In this respect Eckhart and all other thinkers of Eckhart's pattern go on disregarding rules of logic or linguistics. The point is that linguists or logicians are to abandon their limited way of studying facts of experience so that they can analyze the facts themselves and make language amenable to what they discover there. As long as they take up language first and try to adjust all

[14] Evans, p. 228.

human experiences to the requirements of language instead of the opposite, they will have their problems unsolved.

Eckhart further writes:

> The just man serves neither God nor creature: he is free; and the more he is just the more he is free and the more he is freedom itself. Nothing created is free. While there is aught above me, excepting God himself, it must constrain me, however [great]; even love and knowledge, so far as it is creature and not actually God, confines me with its limits.[15]

Let us first see what linguistics would say about this statement. Its reasoning may run something like this: "When Eckhart expresses himself as 'a free man' he is irresistible and wonderful, but he still recognizes God as he confesses 'excepting God.' Why, we may ask, has he to make the exception of God instead of asserting his absolute freedom above all things small and great? If he has to consider God, he cannot be so free as he claims to be?" These objections hold good indeed so far as our logical analysis does not extend beyond language and its values. But one who has an Eckhartian experience will very well understand what he really means. And what he means is this: a man is free only when he is in God, with God, for God; and this is not the condition of freedom, for when he is in God he is freedom itself; he is free when he realizes that he is actually himself by forswearing that he is in God and absolutely free. Says Eckhart:

> I was thinking lately: that *I am a man* belongs to

[15] *Ibid.*, p. 204.

The Basis of Buddhist Philosophy

other men in common with myself; I see and hear and eat and drink like any other animal; but that *I am* belongs to no one but myself, not to man nor angel, no, nor yet to God excepting in so far as I am one with him.[16]

In the latter part of the same sermon, Eckhart adds: "*Ego*, the word 'I,' is proper to none but to God himself in his sameness." This "I" is evidently referred to in another sermon entitled "The Castle of the Soul" as "a spark," "a spiritual light."

From time to time I tell of the one power in the soul which alone is free. Sometimes I have called it the tabernacle of the soul; sometimes a spiritual light, anon I say it is a spark. But now I say: it is neither this nor that. Yet it is somewhat: somewhat more exalted over this and that than the heavens are above the earth. . . . It is of all names free, of all forms void: exempt and free as God is in himself.[17]

Our language is the product of a world of numbers and individuals of yesterdays and todays and tomorrows, and is most usefully applicable to this world (*loka*). But our experiences have it that our world extends beyond that (*loka*), that there is another called by Buddhists a "transcendental world" (*lokauttara*) and that when language is forced to be used for things of this world, *lokottara*, it becomes warped and assumes all kinds of crookedness: oxymora, paradoxes, contradictions, contortions, absurdities, oddities, ambiguities, and irrationalities. Language itself is not to be blamed for it. It is we

[16] *Ibid.*, p. 204.
[17] *Ibid.*, p. 37.

ourselves who, ignorant of its proper functions, try to apply it to that for which it was never intended. More than this, we make fools of ourselves by denying the reality of a transcendental world (*lokottara*).

Let us see how impossible it is to bring a transcendental world or an "inner power" onto the level of linguistic manageability.

There is something, transcending the soul's created nature, not accessible to creature, non-existent; no angel has gotten it for his is a clear[18] nature, and clear and overt things have no concern with this. It is akin to Deity, intrinsically one, having naught in common with naught. Many a priest finds it a baffling thing. It is one; rather unnamed than named, rather unknown than known. If thou couldst naught thyself an instant, less than an instant, I should say, all that this is in itself would belong to thee. But while thou dost mind thyself at all thou knowest no more of God than my mouth does of colour or my eye of taste: so little thou knowest, thou discernest, what God is.[19]

What "a baffling thing" this "something" or "somewhat" is! But it is no doubt a light and if you can get a glimpse into it even "less than an instant" you will be master of yourself. Plato describes the light in the following words: It is "a light which is not in this world; not in the world and not out of the world; not in time nor in eternity; it has neither in nor out." [20] Linguistically considered, how could

---

[18] "*Ein luter wesen*" in German. *Luter* means "intellectually or analytically clear and distinct," opposed to what may be called "metaphysically indefinite."
[19] Evans, pp. 204–5.
[20] Quoted by Eckhart, Evans, p. 205.

a thing be said to be "neither in the world nor in out-of-the-world"? Nothing can be more absurd than this. But, as Eckhart says (Evans, p. 227), when we transcend time (*zit*), body (*liplicheit*), and multiplicity (*manicvaltikeit*),[21] we reach God, and these three things are the very principle of linguistics. No wonder that when things of the *lokottara* try to find their expression through language, the latter shows every trace of its shortcomings. This is the reason why Zen Buddhism strives to avoid the use of language and quite frequently denounces our shortsightedness in this respect. Zen does not object to language just for the sake of opposition, it simply realizes that there is a field in which our words fail to communicate events taking place there. One of the statements Zen is always ready to make is: "No depending on words." Yengo, commentator of the *Hekigan-shu* ("Blue Rock Collection"), a work of the Sung dynasty, thus remarks:

Bodhidharma observing that the Chinese minds are matured enough to accept teachings of Mahāyāna Buddhism came over here [China] via the southern route and started to prepare the people for "the transmission of mind-seal." He said, "I am not going to build up a system of thought which depends on letters or words. I want straightforwardly to direct you to the Mind itself and thereby to see into the Buddha-nature and attain Buddhahood. When Zen is understood in this way, we shall be able to attain freedom. Let us not therefore follow the way of letters of any kind, let us take hold of Reality in its nakedness. To the question of Wu the Emperor of the Liang, Bodhidharma simply answered, "I do not

[21] Pfeiffer, p. 296.

know, your Majesty!" When Eka, who became the second patriarch of Zen in China, confessed that he could not locate the Mind, Bodhidharma exclaimed, "There, I have your mind pacified!" In all these situations which confronted him, Bodhidharma just faced them without hesitation, with no prepared answers concocted beforehand, he had nothing premeditated or deliberately schematized in his concept-filled mind. With one swing of the sword he cut asunder every obstacle that lay in our way, thereby releasing us from the fetters of linguistic discrimination. We are now no more to be troubled with right and wrong, gain and loss.[22]

The following *mondo*[23] will demonstrate how free Zen is in dealing, for instance, with the ultimate problem of being:

A monk asked Daizui Hoshin of the T'ang dynasty: "I am told that at the end of the universe a great fire takes place and everything is destroyed. May I ask you whether or not, 'this' also shares the fate?"

Daizui replied, "Yes, it does."

The monk went on, "If this is the case, it must be said that 'this' follows others."

Daizui: "Yes, it does."

The same question was later asked of another master whose name was Shū. Shū the master answered, "No, it does not." When he was asked "Why not?" the master replied, "Because it identifies itself with the whole universe."

From the logical linguistic point of view the two

[22] A more or less modernized interpretation given to Yengo's terse and loosely knit Chinese.
[23] Literally, "question and answer."

The Basis of Buddhist Philosophy

Zen masters defy each other and there is no way to effect a reconciliation. One says "yes" while the other says "no." As long as the "no" means an unqualified negation and the "yes" an unqualified affirmation, there is no bridge between the two. And if this is the case, as apparently it is, how can Zen permit the contradiction and continue the claim for its consistent teaching, one may ask. But Zen would serenely go its own way without at all heeding such a criticism. Because Zen's first concern is about its experience and not its modes of expression. The latter allow a great deal of variation, including paradoxes, contradictions, and ambiguities. According to Zen, the question of "is-ness" (*isticheit*) is settled only by innerly experiencing it and not by merely arguing about it or by linguistically appealing to dialectical subtleties. Those who have a genuine Zen experience will all at once recognize in spite of superficial discrepancies what is true and what is not.

6

Before I come to another utterance to be ascribed to Buddha at the time of his enlightenment-experience I cannot refrain from considering the problem of time. This is also closely related to linguistics and the Eckhartian treatment of the creation-myth. As Augustine confesses, we can also say that God "mocks at man" when the question of time confronts us. It is one of the subjects of discourse we must "familiarly and knowingly" take up. "And we understand when we speak of it, we understand also when we hear it spoken of by another."

What then is time? If no one asks me, I know; if I wish to explain it to one who asketh, I know not: yet I say boldly that if nothing passes away, time past were not; and if nothing were coming, a time to come were not; and if nothing were, time present were not. Those two times then, past and to come, how are they, seeing the past now is not, and that to come is not yet? But the present, should it always be present, and never pass into time past, verily it should not be time, but eternity. If time present (if it is not time) only cometh into existence, because it passeth into time past, how can we say that either this is, whose cause of being is, that it shall not be; so, merely, that we cannot truly say that time is, but because it is tending not to be? [24]

Time is really an eternally puzzling problem, especially when it is handled at the level of linguistics. As far as linguistics is concerned, the best way to approach the question will be as Eckhart suggests, to consider human beings born "between one and two. The one is eternity, ever alone and without variation. The two is time, changing and given to multiplication." [25] Following this line of thought, Eckhart goes on to say in another sermon on "poverty":

... therefore, I am my own first cause, both of my eternal being and of my temporal being. To this end I was born, and by virtue of my birth being eternal, I shall never die. It is of the nature of this eternal birth that I *have been* eternally, that I *am* now, and *shall be*

[24] St. Augustine, *Confessions*, Book XI, 14.
[25] Evans, p. 134.

The Basis of Buddhist Philosophy

forever. What I am as a temporal creature is to die and come to nothingness, for it came with time and so with time it will pass away. In my eternal birth, however, everything was begotten. I was my own first cause as well as the first cause of everything else. If I had willed it, neither I nor the world would have come to be! If I had not been, there would have been no god. There is, however, no need to understand this.[26]

Whatever Eckhart might have meant by the statement, "There is, however, no need to understand this," it is impossible from the purely linguistic point of view "to understand" the interpenetration or interfusion of time and eternity as described here. Primarily the two concepts, time and eternity, are irreconcilable, and however much dialectical skill one may employ, they can never be brought peacefully together. Eckhart and all other thinkers and non-thinkers may try all their arts to convince us of "the truth" but as long as we are on this side of the stream, we cannot be expected to understand it. This is perhaps what Eckhart means when he says there is no necessity for achieving the impossible. What then does he wish us to do? What he wishes is to turn away from linguistics, to shake off the shackles of "time and matter and multiplicity" and to plunge right into the abyss of nameless nothingness. For it is at the very moment of the plunge that the experience of enlightenment takes place and the understanding comes upon us. "I am that I was and that I shall remain now and forever. Then I receive an impulse which carries me above all angels. In this impulse I conceive such passing riches that I am

[26] Blakney, p. 231.

not content with God as being God, as being all his godly works, for in this breaking-through I find that God and I are both the same. Then I am what I was, I neither wax nor wane, for I am the motionless cause that is moving all things." [27]

When this—regarding my self—is understood, we shall also be led to understand what Augustine says about God: "God is doing today all that shall be done in the thousands upon thousands of years of the future—if the world is to last so long—and that God is still doing today all he ever did in the many thousands of years of past." [28]

Now, both Eckhart and Augustine ask, "What can I do about it if anyone does not understand that?" or "If these words are misconstrued, what can one who puts their right construction on them do about it?" [29] To this, Eckhart answers consolingly: "If anyone does not understand this discourse, let him not worry about that, for if he does not find this truth in himself he cannot understand what I have said— for it is a discovered truth[30] which comes immediately from the heart of God." [31] The only thing that is left for us to do will be to follow Eckhart's advice and pray to God: "That we all may so live as to experience it eternally, may God help us! Amen."

The Zen way of treating the problem of time will

[27] Evans, p. 221.
[28] Blakney, p. 72.
[29] *Ibid.*, pp. 72, 73.
[30] "*Ein unbedahtiu warheit*" (Pfeiffer, p. 284), that is, a truth not premeditated but spontaneously coming to mind.
[31] Blakney, p. 232.

The Basis of Buddhist Philosophy

be partly glimpsed from the following story which contrasts significantly with the linguistic analysis:

Tokusan (790–865), on his way to Taisan, felt hungry and tired and stopped at a roadside teahouse and asked for refreshments. The old woman who kept the house, finding that Tokusan was a great student of the *Diamond Sūtra,* said: "I have a question to ask you; if you can answer it I will serve you refreshments for nothing, but if you fail you have to go somewhere else for them." As Tokusan agreed, the woman proposed this: "In the *Diamond Sūtra* we read that 'The past mind is unattainable; the present mind is unattainable; the future mind is unattainable'; and so, what mind do you wish to punctuate?" (Refreshments are known in Chinese as *t'ien-hsin* or *ten-jin* in Japanese, meaning "punctuating the mind," hence the question.) Tokusan was altogether nonplused and did not know how to answer. He had to go without anything to eat.[32]

Taking refreshments takes place in time. Out of time there is no taking of anything. The old teahouse keeper now asks the traveling monk what time he will use for recuperating from fatigue when, according to the *Sūtra,* no time, past or future or present, is "obtainable." When there is no time how can one accomplish anything? As far as thought is concerned, that is, where language is supreme, no movement of any sort is possible in this life, and yet the strangest fact is that we keep on living in the fullest sense of the term. The old lady is not a

---

[32] *Studies in Zen,* p. 126.

metaphysician, nor is she at all interested in metaphysics. But when she saw how inextricably the young man was involved in verbalism and in its intricate complexities she wanted to rescue him, hence the question. And sure enough he never thought of this possibility. Finding himself in the midst of contradiction, he knew not how to clear himself out of the trap which was of his own construction. He had to go without his refreshments.

Zen is very much interested in the problem of the absolute now-moment, but its interest is more along the practical line and not in its dialectics. Therefore, as in Ummon's "sermon" on "the fifteenth day of the month," the Zen masters want to have us "say a word" (*ikku* or *i-chü* in Chinese). The saying is not necessarily uttering any sound, it is acting of some kind. In Eckhart's term, the trick is to insert "the soul's day" into "God's day" (Blakney, p. 212). God's day is characterized as containing all time in the present now-moment [*in eime gegenwürtigen nu*].[33] To God, "a day, whether six thousand years ago, is just as near to the present as yesterday." We, of another kind of day where yesterday is yesterday and one thousand years is one thousand times more than a year, either to the past or to the future, cannot have God's day operative in our everyday living. But if we do not somehow succeed in making "was" or "will be" turn into "is," we cannot have peace of mind, we cannot escape from dread, which is a topic current among existentially minded modern men. They must somehow

[33] Pfeiffer, p. 265.

The Basis of Buddhist Philosophy

have "the refreshment" served. To have and yet not to have must really be the cause for worry and anxiety.

I will quote a Zen sermon[34] to show how it differs from those sermons given by Eckhart, though it treats the same subject of time and eternity and the basic ideas do not differ so widely as they superficially appear. The Zen sermon was given by Daitō the national teacher (1282–1337) of the fourteenth century who was the first abbot of the Daitoku monastery, Kyoto. A Zen sermon generally begins with a *mondo* between the master and one of the disciples when the sermon proper is very short consisting of about a dozen lines rhetorically composed. The occasion was a New Year's Eve. When Daitō the master appeared in the Dharma Hall, a disciple came forward and proposed the following question:

"The new does not know that the old is already gone while the old does not know that the new is already come. The new and the old have not made acquaintance with each other. Thus they stand in opposition all over the world. Is this the state of affairs we greet on all side?"

The master said, "All over the universe." [35]

The monk continued, "When the world has not yet come into existence,[36] how do we find a passage to it?"

The master: "We fold the hands before the monastery gate."

---

[34] From the *Sayings of Daitō Kokushi*.
[35] The original literally has, "The thirty-three heavens and the twenty-eight constellations."
[36] Literally, "When the firework has not yet been cracked."

The monk: "Anything further?"

The master: "We burn incense at the Buddha Hall."

The monk: "I understand that, anciently, Hokuzen roasted the big white bullock[37] that used to roam at the monastery courtyard and gave a feast to his monks to celebrate this memorable occasion. I wonder what kind of feast we are going to get this New Year's Eve?"

The master: "When you chew it fine, it tastes sweeter than honey."

The monk: "In this case we of the Brotherhood will appreciate your generosity."

As the monk bowing began to step back the master said, "What a fine golden-haired lion!" [38]

The master now gives his regular sermon: "The old year passes away this evening. Let things go that are to go and grow old. The new year is ushered in at this dawn. Let things come that are to come and be renewed. The new and the old are intermingled in every possible way, and each of us enjoys himself as he pleases. Causes and effects go on in time-sequence, and everywhere activities in every form manifest themselves freely and autonomously. Thus we observe the peak of Mount Ryūho magnificently towering to the sky, while the monastery gate opens to a field limitlessly expanding. This is not altogether due to the peaceful time alone we now enjoy under the wisely governing reign. It is in the order of things that the spirit of universal friendliness

---

[37] "The white bullock" is the symbol for the Highest Reality.
[38] This is a more or less ironical remark on the part of the master.

pervade all around us. At this moment what lines shall I quote for your edification?"

The master struck the seat with his *hossu* and said, "The December snows filling up to the horizon make all things look white, while the spring winds blowing against the doors are still severely cold."

According to the lunar calendar, the thirtieth day of the twelfth month (December) is the end of winter and as soon as twelve o'clock is struck spring is ushered in. Hence Daitō's reference to December snows and spring winds. They are both there: the winter snows do not melt away when spring starts. The spring breeze passes over the same old winter snows. The old and the new are mingled. The past and the present are fused. The imaginary line of season exists only in human language. We for some practical purposes distinguish seasons. When this is once done one season must definitely start in such and such time of the year. While the snow lies white, it does not make haste to greet spring and the wind does not wait for winter to make way for it. The old continues on to the new and the new is ready to join the old. Zen's absolute present is probably not so inaccessible as Eckhart's.

7

It is now time, after these lengthy excursions, to come back to the original topic and see if we cannot get once more into the subjective approach to Buddha's experience of enlightenment. The experience cannot merely be designated as a kind of feeling and thus done away with as if this designation exhausted all the contents of enlightenment. For, as I understand it, the enlightenment cannot be said to be devoid of any noetic elements which yield to a certain extent to a linguistic and intellectual treatment. The feeling of enlightenment has something profoundly fundamental and gives one a sense of absolute certainty and finality which is lacking in the ordinary kind of feeling we generally have. A feeling may occasionally give one the sense of exaltation and self-assurance, but this will after a while pass away and may leave no permanent effect on the being of one who has the experience. The enlightenment feeling on the other hand affects the whole personality, influencing his attitude toward life and the world not only morally and spiritually but in his metaphysical interpretation of existence as a whole. Buddha's experience was not just a matter of feeling which moves on the periphery of consciousness, but something awakened in the deepest recesses of a human being. In this sense only is his utterance recorded in the *Vinaya* and the *Majjhima Nikāya* and elsewhere to be understood. In the *gāthā* already quoted above from *The Dhammapada* (vv. 153, 154), something similar to the one below is noticeable, but the positive and dynamic aspect comes forward more strongly and conspicuously in the following:[39]

> I have conquered and I know all,
> I am enlightened quite by myself and have
>   none as teacher.

[39] *The Vinaya*, I, p. 8. *The Majjhima Nikāya* (translated by Lord Chalmers, published by Oxford University Press), 26, p. 12.

The Basis of Buddhist Philosophy

> There is no one that is the same as I in the whole world where there are many deities.
> I am the one who is really worth,
> I am the most supreme teacher.
> I am the only one who is fully enlightened.
> I am tranquillized.
> I am now in Nirvana.[40]

This victory song is expressive of the supreme moment of the enlightenment-experience which Buddha had. In the first verse depicting the discovery of the *gahakāraka* (house-builder) and the demolition of his handiwork, we see the negative aspect of Buddha's experience, while in the second one dealing with the exalted feeling of victory, the realization of the highest knowledge (*prajñā*) and the consciousness of one's own value as he is, we see its positive aspect coming out in full view.

The consciousness of conquest such as was awakened in the mind of Buddha at the time of enlightenment cannot be regarded as the product of a self-conceit which is often cherished by minds tarnished with schizophrenia and the wielders of political or military powers. With him however whose ego-centered desires have been shattered to pieces the consciousness of victory rises from the deepest sources of being. So the feeling of conquest is not the out-

---

[40] It will be interesting to note that we have another *gāthā* in *The Dhammapada*, v. 353, which also echoes the same sentiment as the one here quoted from another source. It is possible that they are from one and the same original source. *The Dhammapada* one runs thus:

> I have conquered all, I know all, in all conditions of life I am free from taint. I have left all, and through the destruction of thirst I am free. Having by myself attained specific knowledge, to whom can I point as my teacher?

come of a struggle of powers belonging to the low level of existence. The enlightenment-experience is the manifestation of a higher power, a higher insight, a higher unification. It is beyond the sphere of relative consciousness which is the battleground for forces belonging to the same order. One force may temporarily proclaim its victory over another, but this kind of victory is sure before long to be superseded by another. This is in the nature of our relative consciousness. Enlightenment is the experience a man can have only when a higher realm of unification is revealed, that is, when the most fundamental basis of identification is reached.

The enlightenment-experience, therefore, is the one which we can have only when we have climbed up to the highest peak from which we can survey the whole field of Reality. Or we can say that it is the experience which is attained only when we have touched the very bedrock which sustains the entire system of multiple worlds. Here is the consciousness of intensive quantity to which nothing more could be added. All is fulfilled, satisfied; everything here appears to it such as it is; in short, it is a state of absolute Suchness, of absolute Emptiness which is absolute fullness.

Buddhist philosophy, therefore, is the philosophy of Suchness, or philosophy of Emptiness, or philosophy of Self-identity. It starts from the absolute present which is pure experience, an experience in which there is yet no differentiation of subject and object, and yet which is not a state of sheer nothingness. The experience is variously designated: in Japanese it is *sono-mama;* in Chinese it is *chih mo,*

The Basis of Buddhist Philosophy

sometimes *tzu-jan fa-erh* (Japanese: *jinen hōni*); there are many technical names for it, each denoting its specific features or characters as it is viewed in various relationships.

In fact, this Suchness, or "is-ness" (*isticheit*) in Eckhart's terminology, defies all characterization or denotation. No words can express what it is, but as words are the only instrument given us human beings to communicate our thought, we have to use words, with this caution: Nothing is available for our purpose; to say "not available" (*anupalabda* in Sanskrit and *pu k'o tê* in Chinese) is not to the point either. Nothing is acceptable. To say it is, is already negating itself. Suchness transcends everything, it has no moorings. No concepts can reach it, no understanding can grasp it. Therefore, it is called pure experience.

In pure experience there is no division between "ought" and "is," between form and matter or content, and therefore there is no judgment in it yet. There is the Christ who says "I am before Abraham was," or God who has not yet uttered his fiat. This is Buddha who, according to *The Dhammapada* (179), is the *anantagocara* ("one whose limits are infinite"), the *apada* ("the pathless"), whose conquest can never be conquered again and into whose conquest nobody in this world can enter, and who is where there is no track leading to it. If it were a Zen master, he would demand that you show your face, however ugly it might be, which you have even before your birth into this world of multiplicities.

The Buddhist philosophy of Suchness thus starts with what is most primarily given to our consciousness—which I have called pure experience. But, in point of fact, to say "pure experience" is to commit oneself to something already posited somewhere, and thus it ceases to be pure. *The Dhammapada* reflects this thought when it designates the starting point of Buddhist philosophy as trackless (*apada*), unboundable (*anantagocara*), abodeless (*aniketa*), empty (*śuñña*), formless (*animitta*), delivered (*vimokkha*). In psychological terms, it is described thus: sorrowless (*vippamutta*), released on all sides (*sabbaganthappahīna*), fearless (*asantāsin*), without craving (*vītataṇha*). These psychological terms are apt to be very much misunderstood because they point to negativism when superficially and linguistically interpreted. But I will not dwell upon this here.

One thing that must be noted in this connection is that pure experience is not pure passivity. In fact there is nothing we can call pure passivity. This does not make sense and does not lead us anywhere. As long as passivity is also an experience, there must be one who experiences passivity. This one, this experiencer, is an actor. Not only is he an actor, but he is a knower, for he is conscious of experiencing. Pure experience is not an abstraction or a state of passivity. It is very much active, and creative. Eckhart voices this idea when he states: "In this sense thy unknowing is not a defect but thy chief perfection, and suffering thy highest activity. Kill thy activities and still thy faculties if thou wouldst realize this birth in thee." [41]

[41] Evans, p. 14. "This birth" in this sermon means "the newborn Being" or "the child of man turned into the child of God." It also means "hearing of the Word" which is revealed to "one who knows aright in unknowing."

The Basis of Buddhist Philosophy 79

Another thing I should like to emphasize in this *gāthā* of conquest is that Buddha calls himself "all-conqueror" and also "all-knower," showing that his victory is absolute and that his knowledge is not at all fragmentary. He is omniscient as well as omnipotent. His experience has something noetic and at the same time something conative or affective, reflecting the nature of Reality itself which consists in *prajñā* and *karuṇā*. As regards *prajñā*, which is sometimes translated as "transcendental wisdom," I have written about it elsewhere. Therefore I shall speak here about *karuṇā*. *Karuṇā* corresponds to love. It is like the sands on the Ganges: they are trampled by all kinds of beings: by elephants, by lions, by asses, by human beings, but they do not make any complaints. They are again soiled by all kinds of filth scattered by all kinds of animals, but they just suffer them all and never utter a word of ill-will. Eckhart would declare the sands on the Ganges to be "just" (*gerecht*), because "the just have no will at all: whatever God wishes it is all one to them, however great the discomfort may be." [42]

The just are so firmly devoted to justice and so wholly selfless that whatever they do, they regard neither the pains of hell nor the joys of heaven.... To the just person, nothing is so hard to bear or so painful as anything opposed to justice, that is to say, as not feeling impartially the same about everything that happens.[43]

"Justice" savors a great deal of legalism contrary to the idea of love. But when, as Eckhart interprets it, justice is considered from the affective point of

[42] Blakney, p. 179.
[43] *Ibid*.

view as meaning "impartiality," "sameness," "universality," or "all-embracing," it begins to approach the Buddhist idea of *karuṇā*. I may add that Mahāyāna Buddhism further developed the idea of *karuṇā* into that of *praṇidhāna* or *pūrvapraṇidhāna* and made each one of the Bodhisattvas an incarnation of a certain number of *praṇidhāna*, for example, Amitābha has forty-eight *praṇidhāna*, Samantabhadra has ten, and Kṣitigarbha also has ten. *Praṇidhāna* is generally translated as "vow" or "fervent wish" or "prayer," or simply "the will," but these English terms do not convey the full meaning of the Sanskrit as it is used in the Mahāyāna. Roughly speaking, we may interpret *praṇidhāna* as love specified or itemized or particularized and made applicable to each practical situation in which we may find ourselves in the course of an individual life. Amitābha has his Pure Land where he wants us to be born; Mañjuśrī is the Bodhisattva of *prajñā* and whoever comes to him will be rewarded with an amount of transcendental wisdom.

This being the case, we will see that "the destruction of desires or cravings (*taṇhānam khayam*) so much emphasized in the teaching of earlier Buddhism is not to be understood negativistically. The Buddhist training consists in transforming *tṛiṣṇā* (*taṇhā*) into *karuṇā*, ego-centered love into something universal, eros into agape.

When Jōshu (778–897) was asked, "Could Buddha cherish any desires (*kleśa*)?" he answered, "Yes, he decidedly has." The questioner demanded, "How could that be?" The master replied, "His desire is to save the whole universe."

The Basis of Buddhist Philosophy

One day Jōshu had another visitor who asked, "I hear so much of the stone bridge reputed to be on one of the sites in your monastery grounds. But as I see it, it is no more than an old log. How is that?"

The master said, "You see the log and don't see the stone bridge."

"What is the stone bridge, then?" the visitor demanded.

The master's answer was, "It permits horses to pass and also asses to pass."

Someone's *praṇidhāna* is too rickety for safe crossing whereas the other's is strong and broad, allowing anything to pass over it safely. Let *taṇhā* be destroyed but we must not forget that it has another root which reaches the very ground of being. The enlightenment-experience must realize that, though ordinarily Buddhists are more or less neglectful in bringing out the *karuṇā* aspect of the experience. This is due to their being too anxious and therefore too much in a hurry to destroy all the obstacles lying on the way to enlightenment, for they know that when this is accomplished what is to come therefrom is left to itself as it knows full well how to take care of it. When the devastating fire is extinguished the forest will not wait for any external help but will resume its biological functions by itself. When a man is shot by a poisonous arrow the first thing to do is to remove it before it is embedded too deeply into the flesh. When this is done the body will heal the wound by its own power of vitality. So with human passions, the first work is to destroy their root of ignorance and egoism. When this is thoroughly accomplished, the Buddha-nature which consists in *prajñā* and *karuṇā* will start its native operation. The principle of Suchness is not static, it is full of dynamic forces.

Let me finish this study on Buddha's enlightenment-experience by quoting another from Daitō Kokushi's *Sayings*. In Japan and China Buddha is recorded to have attained his enlightenment on the morning of December the eighth. After a period of deeply absorbed meditation he happened to look skyward and there he noticed the morning star shining brightly. This at once caused something like a flash of lightning to pass through his consciousness, which put a final stop to his quest for the truth. He felt as if all the burden which he had been carrying dropped off his shoulders and a long sigh of relief came out of his being. The Zen Buddhists are specially mindful of remembering this event and a commemoration always takes place on December the eighth.

When Daitō the national teacher appeared in the Dharma Hall, a monk came out of the rank and started a series of questions:

"The record has it that the Bodhisattva attained enlightenment today and that since then he is known as Tathāgata. But what took place in his mind when he saw the morning star? What did he understand?"

The teacher replied: "Thoroughly clean! Utterly blank!"

The monk: "Even when there is one speck of dust in your eye, does it not make you see all kinds of imaginary flowers in the air?"

The Basis of Buddhist Philosophy

Teacher: "Don't be too talkative!"

Monk: "Would you approve my going on along this line?"

Teacher: "You ask my staff which knows better than I."

Monk: "Things are going on today just as they did in the past; why refuse to approve my case?"

Teacher: "It's only because there still is a dualism."

Monk: "If not for this remark I should surely have missed your point."

Teacher: "You beat me with your argument."

Monk: "When one is genuinely sincere one has nothing to be ashamed of."

Teacher: "There are many like you." [44]

Daitō the national teacher then gave a short sermon:

> The moon is clear and serene,
> The stars are shining bright.
> No Śākyamuni is here,
> And whose enlightenment are we talking about?

He now held his staff up straight and declared: "It is piling filth over filth."

---

[44] These questions and answers may sound somewhat enigmatic to most of our readers. A full explanation will be given elsewhere on a more opportune occasion.

chapter 3

"A Little Point" and Satori

I

Meister Eckhart is quoted in Inge's *Mysticism in Religion* (p. 39):

The union of the soul with God is far more inward than that of the soul and body. . . . Now, I might ask, how stands it with the soul that is lost in God? Does the soul find herself or not? To this I will answer as it appears to me, that the soul finds herself in the point where every rational being understands itself with itself. Although it sinks in the eternity of the divine essence, yet it can never reach the ground. Therefore God has left a little point wherein the soul turns back upon itself and finds itself, and knows itself to be a creature.

An interesting controversy arises regarding "a little point" referred to in this passage from Eckhart. I do not know where Dean Inge found the passage. It would be desirable to know the entire context where it occurs, if we would discuss the point fully, but we can somehow proceed according to the extent of our general knowledge of Eckhartian philosophy. One person insists that Eckhart here tells us about the human impossibility of reaching the ground of Reality or "the inmost core [*grund*] of the divine nature." According to this interpretation, there is an impassable gap between "every rational being" and "the eternity of the divine essence"; God provides us, therefore, with "a little point" whereby we rational beings are made to turn upon ourselves and realize that we are after all finite creatures and barred forever from sinking into "the core of God" or "the essence of God."

The other person's way of thinking runs along the following line: Judging from the whole trend of Eckhart's ideas as expressed in his sermons, he does not necessarily mean here that the gap between the divine ground and ourselves is absolutely impassable; on the contrary, he implies that he himself crossed the gap and came back to this side of rationality. This person will insist that if Eckhart did not cross the impassable himself how could he say that "God has left a little point" as if he were God himself? Or, logically speaking, when Eckhart says that there is a gap and that the gap is impassable, he must have already been there and seen the gap and actually surveyed it and found it impassable.

In our relative way of thinking, the finite is sharply differentiated from the infinite, they cannot be made one, there is no way of unifying them. But when we analyze the concept closer, we find that one implies or participates in the other and that because of this implication or participation the one becomes separable from the other in our thought. Disintegration is possible because of integration, and vice versa. It is in this sense that the finite is infinite and the infinite is finite.

But here is a subtle point over which we must take care not to slip: When we say that the finite is infinite, it does not mean that the finite as it stands relatively as finite is infinite and so with the infinite; they pass into each other and become one

"A Little Point" and Satori

when all ideas of relativity are wiped away. But we must be quite cautious not to undertake this wiping away on the relative plane, for in this case there will have to be another wiping away and this may go on eternally. This is where many intellectuals stumble and become victims of their own cleverness.

When they talk about an impassable gap and a point wherein we are made to turn back, they forget that by this very talk they are already crossing the impassable and find themselves on the other side. It is due to their discriminating habit of thought that the impassable is always left on the other side while they are actually there. We are possessed of the habit of looking at Reality by dividing it into two; even when we have in all actuality the thing we spend time in discussing and then finally come to the conclusion that we have it not. It is all due to the human habit of splitting one solid Reality into two, and the result is that my "have" is no "have" and my "have not" is no "have not." While we are actually passing, we insist that the gap is impassable.

When Eckhart says that "God has left a little point," this is, according to my understanding, to remind us of the fact that we are all finite beings, that is, "creatures," and therefore that as such we "can never reach the ground." But inasmuch as we are "sinking in the eternity of the divine essence" we are already on the ground. This is where we are God himself. It is only when we *see* the "little point" left by God that we return to ourselves and know that we are creatures. This *seeing* is splitting and all forms of bifurcation take place and we are no more

God, we are no more, as Eckhart says, "One to one, one from One, one in One and the One in one, eternally." This is "where time comes in and all the properties of things which belong to time—existing beside the timeless."[1] We all make time sit beside timelessness. But why not time in timelessness and timelessness in time?

2

"A little point" left by God corresponds to what Zen Buddhists would call *satori*. When we strike this point we have a *satori*. To have a *satori* means to be standing at Eckhart's "point" where we can look in two directions: God-way and creature-way. Expressed in another form, the finite is infinite and the infinite is finite. This "little point" is full of significance and I am sure Eckhart had a *satori*.

Eckhart's "little point" is the eye, that is to say, "The eye by which I see God is the same as the eye by which God sees me. My eye and God's eye are one and the same—one in seeing, one in knowing, and one in loving."[2] Eckhart says: "If my eye is to distinguish colors, it must first be free from any color impressions. If I see blue or white, the seeing of my eyes is identical with what is seen." If the seeing is the seen and the seen is the seeing, the "sinking in the eternity of the divine essence" is the "reaching the ground," for there cannot be any "ground" which "is beside the timeless," the ground is the divine essence, and the sinking is reaching.

[1] Blakney, p. 81.
[2] *Ibid.*, p. 206.

"A Little Point" and Satori 89

What makes us however think that Eckhart really advocates the doctrine of impassableness is that here he seems to remind us of our being creatures more than of our being one with God or our coming from the core (or *grund*) of the divine nature. The "little point" here referred to is made to turn us around back to our finite creatureliness, but the fact is that the point can readily be made to turn the other way leading us straight to the Godhead. Eckhart calls the one who can achieve this wonder the aristocrat (*Edel*), and defines him:

> So I say that the aristocrat is one who derives his being, his life, and his happiness from God alone, with God and in God and not at all from his knowledge, perfection, or love of God, or any such thing. Thus our Lord says very well that life eternal is to know God as the only true God and not that it is knowledge that God may be known.[3]

According to this, Eckhart distinguishes two kinds of knowledge: one is to know God as the only true God and the other is to know God through knowledge about him. The second kind is "twilight knowledge in which creation is perceived by clearly distinguished ideas"; while the first kind is "daybreak knowledge" where "creatures are known in God," and "in which creatures are perceived without distinctions, all ideas being rejected, all comparisons done away in that One that God himself is."[4] Cannot this kind of knowledge be called also the knowledge of impassableness? Is this not what a

[3] *Ibid.*, p. 80.
[4] *Ibid.*, p. 79.

man gets when the "little point" makes him turn around God-way in which all creatures, all distinctions, all comparisons, all ideas are done away with, leaving God to be in himself and with himself?

Eckhart states in "The Aristocrat," from which the above quotations are called:

> Neither the One, nor being, nor God, nor rest, nor blessedness, nor satisfaction is to be found where distinctions are. Be therefore that One so that you may find God. And of course, if you are wholly that One, you shall remain so, even where distinctions are. Different things will all be parts of that One to you and will no longer stand in your way.[5]

Where distinctions are you cannot find "the One" or "Being," but when you are "that One," "wholly that One," all distinctions or all different things may be left as they are and will all be parts of that One and offer you no hindrances, to use Kegon phraseology. To tell the truth, however, distinctions can never remain as distinctions if they were not made "parts of that One," though as far as I am concerned I do not like the term "parts" in connection with the One. "All different things" are not parts but they are the One itself, they are not parts as if they, when put together, would produce the whole. "Parts" is a treacherous term.

Eckhart continues: "The One remains the same One in thousands of thousands of stones as much as in four stones: a thousand times a thousand is just as simple a number as four." This idea of number is really at the bottom of the doctrine of impassable-

[5] *Ibid.*, p. 78.

ness. The idea of distinguishing passable from impassable, or finite from infinite, is derived from the notion of duality, of one divided into two, and these two as standing absolutely against each other, or as contradicting each other, or as irreconcilably excluding each other—which makes it impossible to go over to the other. The One does not belong in the category of number, yet the intellectually strained mind tries to pull it down to its own level. Language is a useful means of communication and expression, but when we try to use it for the deepest experience man can have we trap ourselves and do not know how to extricate ourselves. Eckhart is troubled in the same way as we can see in the following extracts:

> I say that when a man looks at God, he knows it and knows that he is the knower. That is to say, he knows it is God he is looking at and knows that he knows him. Now some people wish it to appear that the flower, the kernel of blessing is this awareness of the spirit, that it is knowing God. For if I have rapture and am unconscious of it, what good would it do and what would it mean? I cannot agree with this position.[6]

According to this, Eckhart is apparently not satisfied with merely knowing God and being conscious of this knowing on the part of the spirit, for he goes on to declare that spiritual blessing consists in absolutely being absorbed in God and not knowing it at all:

> For granting that the soul could not be happy without it [that is, being conscious of its own processes], still its

[6] *Ibid.*, p. 79.

happiness does not consist in that; for the foundation of spiritual blessing is this: that the soul look at God without anything between; here it receives its being and life and draws its essence from the core [*grund*] of God, unconscious of the knowledge-process, or love or anything else. Then it is quite still in the essence of God, not knowing at all where it is, knowing nothing but God.

Evidently here Eckhart thinks that knowing is something between the knower and God, that being conscious of God's presence is not being "quite still in the essence of God," and therefore that there is no foundation here on which spiritual blessing may be established. In this Eckhart is quite right if the knowing of God is to be understood in the way we generally understand knowledge, as issuing from the relationship of subject and object. As he says, "When the soul is aware that it is looking at God, loving him and knowing him, that already is a retrogression, a quick retreat back to the upper level of the natural order of things." [7] To be conscious of knowing God is to know about God if this knowing follows the ordinary way of knowledge-process. But what kind of knowledge does he wish us to understand by the knowledge referred to in the following passage:

For a man must himself be One, seeking unity both in himself and in the One, which means that he must see God and God only. And then he must "return," which is to say, he must have knowledge of God and be conscious of his knowledge.

[7] *Ibid.*, pp. 79-80.

"A Little Point" and Satori

What kind of knowledge does he mean here? In this knowledge is there the division of subject and object? If this is the knowledge of an absolute unity of God and man, what does this "being conscious of his knowledge" imply? When Eckhart then tells us that we "must have knowledge of God" and that we must "return," does this mean that we after all give up "the foundation of spiritual blessing" and retreat to the natural order of things? What difference could there be between "spiritual blessing" and knowledge of absolute oneness? Is the rapture of spiritual bliss preferable to "stepping beyond creatures" or "jumping past creatures" and knowing God? [8]

Eckhart quotes John, 19:12, "A certain nobleman went out into a far country to receive for himself a kingdom, and to return." "The nobleman" means, according to Eckhart, "a person who submits completely to God, giving up all he is and has"; "to go out" means that he "has nothing more to do with vanity . . . to the extent that he is now pure being, goodness and truth"; and then he has "daybreak knowledge in which creatures are perceived without distinctions." But, according to Eckhart, this knowledge is not enough, the nobleman is to be completely free from all forms of knowledge. And then Eckhart continues, "There will be no blessing except a man be conscious of his vision and knowledge of God, but it is not the will of God that I be blessed on that basis. If anyone will have it otherwise, let him do so; I can only pity him."

[8] *Ibid.*, p. 166.

Eckhart is here deeply involved in contradictions. He appraises knowledge, then repudiates it, and finally takes it up again as the thing desired. It is not apparently enough for the nobleman "to go out" and he is advised "to return." In his process of going out and receiving a kingdom, mere knowledge of the oneness of God and himself is no more than knowing *about* God. Such knowledge, it goes without saying, is far from being satisfactory to anyone who in all sincerity seeks God. The soul must "look at God without anything between" and "receive its being and life and draw its essence from the core [*grund*] of God." But when this is accomplished the nobleman is "to return," for he "must have knowledge of God and be conscious of his knowledge." Eckhart seems to be using knowledge in two different senses, one in a relative sense and the other in the absolute. Hence this apparent confusion.

The real fact, however, is that as far as we are human beings we cannot express in words our understanding of Reality in its suchness. When we try to do so we are inevitably involved in a contradiction. Eckhart says, "God's sight and mine are far different—utterly dissimilar."[9] Inasmuch as he could make this statement in regard to God's sight as being utterly dissimilar to our human sight, he must be said to have had certain knowledge of God which enabled him to bring these tidings to the human world from the other shore, from "the inmost core of the divine nature in its solitude."[10] "If I am to see color, I must have that in me which

[9] *Ibid.*, p. 81.
[10] *Ibid.*

is sensitive to color, but I should never see color if I did not have the essence of color already." [11] Unless God was not already with us, in us, we could never know how dissimilar or how similar—which is after all the same thing—God was to us. In this connection Eckhart quotes St. Paul and St. John: "We shall know God as we are known by him" and again, "We shall know God as he is." An image may be "dissimilar" to the object whose image it is, but there is no doubt that it represents the original and to that extent the image must be said to be "similar" to the original. What makes the image an image is the presence in it of the original and as such the image is just as real as the original. The original sees itself in the image as well as in itself. Being in "dissimilarity" must be said to be only in similarity. To realize this is the meaning of "returning."

To quote Eckhart again, "The soul must step beyond or jump past creatures if it is to know God." [12] But to know God is to know oneself as creature. To know God is "to go out" as the Biblical nobleman does according to Eckhart, and his "returning" means knowing oneself as creature by knowing God. When the soul knows God it becomes conscious of its oneness with God and at the same time it realizes how "dissimilar" it is. The "going out" is "returning" and conversely. This circular and contradictory movement characterizes our spiritual experience. A Zen master once produced a staff before the congregation and said: "If you have a staff I will give one to you; if you do not I will

[11] *Ibid.*, p. 168.
[12] *Ibid.*, p. 166.

take it away from you." The giving is the taking away, and the taking away is the giving. Another master later gave his view, saying, "You all throw your staff down!" As long as the staff crosses our way, the question of similarity and dissimilarity, of passableness and impassableness will never be conclusively settled.

3

Eckhart's idea of the "little point" which God left in order to make us turn back to ourselves and realize that we are after all creatures is highly suggestive and full of significance. Most readers are apt to regard such a statement as this as not really touching their own spiritual experience but as something general and impersonal which may be turned to a subject of philosophical discourse. Of course there is no harm in this as long as the statement is understood as reflecting one's personal experience in the matter.

Eckhart's "little point," according to my view, is not just a point which stays stationary. It moves or rather revolves and this movement is taking place all the time. That is to say, the point is a living one and not a dead one. Therefore as soon as we come to this point God may make us turn back toward creatureliness but at the same time he does not forget to remind us of the other side of the point. If the point is stationary and points just one way, we cannot even turn back to ourselves and find ourselves to be creatures. The reason we can turn back is because we can move on and see into the ground (*grund*) of the divine nature. In fact, while going back to our creatureliness we are all the time carrying with us the ground itself, for we cannot leave it behind as if it were something which could be separated from us and left anywhere by the roadside and perhaps picked up by somebody else. Creatureliness and Godliness must go hand in hand; wherever one is found the other is also always there. To leave one behind means killing the other as well as oneself. The "little point" is a kind of axis around which we and God move. This truth will be experienced when a man once actually reaches the point. Then the problem of impassableness no longer remains with him, he will never ask himself whether he can pass on or not. He is what he was. To know the significance of the point one must see it, for God did not leave it where it is in order to make philosophers or theologians argue about its presence so as to help them advance the theories already constructed in their own minds.

Some may say that if the "little point" exists only to make us realize that we are after all creatures, what is the use of looking into the eternity of the divine essence? We all know that we are creatures even before we come to the "little point." This is however no more than mere arguing for the sake of arguing. We must remember that this seeing the "little point" makes the greatest possible difference in the world. We are indeed different creatures, we are not the same creatures any longer after our encounter with the "little point." We are now creatures in God, with God, not creaturely creatures. There are those who think that the "little point" divides us from God forever, and that when we are away from it we have eternally left God on the other side of it. The fact is just the contrary. When we turn back to ourselves after being accosted by the "little point," we have captured everything around there and are carrying it all with us. If things were otherwise we should all find ourselves deeply buried in the emptiness of the Godhead, which means an end to our creatureliness. For the fullness of the Godhead can only be expressed in the creatureliness of all beings.

I do not think it is justifiable to use this "little point" for the support of the doctrine of impassableness. In other places Eckhart gives us statements quite contradictory to the idea of the "little point." For instance, in one of his sermons, "Into the Godhead," [13] we have the following:

As long as the least of creatures absorb your attention, you will see nothing of God, however little that creature may be. Thus, in Book of Love, the soul says: "I have run around looking for him my soul loves and found him not." She found angels and many other things but not him her soul loved, but she goes on to say: "After that, I went a little further and found him my soul loves." It was as if she said: "It was when I stepped beyond creatures that I found my soul's lover." The soul must step beyond or jump past creatures if it is to know God.

This sermon is given under the heading, "A little, and ye see me no more," which means, according to Eckhart, "However small it may be, if anything

[13] *Ibid.*, pp. 165 *et seq.*

adheres to the soul, you cannot see me," that is, God. And: "Every creature seeks to become like God. If there were no search for God, the heavens themselves would not be revolving. If God were not in everything, nature would not function nor would desire be in anything." And this desire is to see God in his naked essence.

If all the shells were removed from the soul and all God's shells could be taken off too, he could give himself directly to the soul without reserve. But as long as the soul's shells are intact—be they ever so slight—the soul cannot see God. If anything, even to the extent of a hairbreadth, came between the body and the soul, there could be no true union of the two. If that is the case with physical things, how much more true it is with spiritual! Thus Boethius says: "If you want to know the straight truth, put away joy and fear, confidence, hope and disappointment." Joy, fear, confidence, hope, and disappointment are all intervening media, all shells. As long as you stick to them and they to you, you shall not see God.

These are all significant and illuminating statements whereby we can look into the core of Eckhart's philosophical thinking. He never wants us to leave the Godhead behind, he just wants us to leave our shells and also asks of God to take off his shells if he has any except those we have put on him. Both we and God are to be naked if there is to be a unification or identity of any sort between the two. To be naked means to be empty, for the two, God and creatures, can join hands only when both stand in the field of Absolute Emptiness (*śūnyatā*), where there is neither light nor shadow.

Let us consider other passages from Eckhart for our own further edification on the subject. The following are from the sermon with the title, "Distinctions Are Lost in God":[14]

> Man's last and highest parting occurs when, for God's sake, he takes leave of God. St. Paul [15] took leave of God for God's sake and gave up all that he might get from God, as well as all he might give—together with every idea of God. In parting with these, he parted with God for God's sake and yet God remained to him as God is in his own nature—not as he is conceived by anyone to be—nor yet as something yet to be achieved—but more as an is-ness (*isticheit*), as God really is. Then he neither gave to God nor received anything from him, for he and God were a unit, that is, pure unity.

Statements like this must have struck Christians of his days as most extraordinary, even as blasphemous, and probably they may still affect present-day Christians in the same way. But from the Buddhist point of view they would not sound in any way strange or singular or astounding. They are rather a routine expression of Buddhist thought. Eckhart however does not stop here, he goes on:

> God gives to all things alike and as they proceed from God they are alike. . . . A flea, to the extent that it is

---

[14] *Ibid.*, pp. 203 *et seq*.

[15] Eckhart quotes St. Paul as saying: "I could wish to be cut off eternally from God for my friends' sake and for God's sake." This I understand corresponds to the King James version, Romans, 9:3, "But I could wish that myself were accursed from Christ for my brethren, my kinsmen according to the flesh." I do not know how this discrepancy takes place between the two quotations as I have no Greek texts with me. Eckhart bases his argument on his Latin text, I believe.

in God, ranks above the highest angel in his own right. Thus in God all things are equal and are God himself. ... In this likeness or identity God takes such delight that he pours his whole nature and being into it. His pleasure is as great, to take a simile, as that of a horse, let loose to run over a green heath where the ground is level and smooth, to gallop as a horse will, as fast as he can over the greensward—for this is a horse's pleasure and expresses his nature. It is so with God. It is his pleasure and rapture to discover identity, because he can always put his whole nature into it—for he is this identity itself.

Is this not a remarkable utterance of spiritual intuition on the part of the author? Here, we see that God, instead of being left behind the "little point," is right out on the greensward with "his whole nature and being" in full display. He keeps nothing in reserve. He gallops like a horse, he sings like a bird, he blooms like the flower, he even dances like a young girl. Living among the conventionally minded tradition-bound medieval Christians, Eckhart must have felt somewhat constrained in his expression and did not go so far as the Zen master would. Otherwise, Eckhart might have had "the wooden horse neigh and the stone man dance," with the same facility as the Zen master.

In one sense, this "little point" may be considered as corresponding to the Buddhist idea of *ichi-nen* (*ekacittakṣāṇa* or *ekakṣāṇa* in Sanskrit and *i-nien* in Chinese). Eckhart's "little point," if I understand it correctly, marks the turning point in the suchness of the Godhead. As long as the Godhead remains in its suchness, that is, in its naked essence, it is Emptiness itself, no sound comes from it, no odor issues from it, it is "above grace, above intelligence, above all desire," [16] it is altogether unapproachable, unattainable, as Buddhist philosophers would say. But because of this "little point" left by it, it comes in contact with creatures by making "the soul turn back to itself and find itself and know itself to be a creature." The time when the soul becomes conscious of its creatureliness is the time also when God becomes aware of his contact with creatures. Or we can say that this is creation. In Sermon 28 we have:

Back in the Womb[17] from which I came, I had no God and merely was, myself. I did not will or desire anything, for I was pure being, a knower of myself by divine truth. The I wanted myself and nothing else. And what I wanted was, and what I was I wanted, and thus I existed untrammelled by God or anything else. But when I parted from my free will and received my created being then I had a God. For before there were creatures, God was not God, but rather he was what he was. When creatures came to be and took on creaturely being, then God was no longer God as he is in himself, but God as he is with creatures.[18]

The Godhead must become God in order to make itself related to creatures. The Biblical God as the creator of the world is no longer God as he was. He created himself as he is, by creating the world. But

[16] *Ibid.*, p. 231.
[17] The original German is *"in miner ersten ursache."* The translator has substituted this term. I do not know whether this is a happy translation or not.
[18] Blakney, p. 228.

"A Little Point" and Satori

even this God is not to be conceived in terms of time. The chronological God is the creation of a relative mind and as such we can say that he is far removed from the Godhead. He is just one of the creatures like ourselves. Eckhart says: "If a flea could have the intelligence by which to search the eternal abyss of divine being, out of which it came, we should say that God together with all that God is could not give fulfillment or satisfaction to the flea!" [19] A chronological God has to have the intelligence of the flea if he wants to delve into the very being of the flea. The rising of this intelligence in the soul, to use Eckhartian terminology, is the positing of the "little point."

---

[19] *Ibid.*, p. 229.

chapter 4

Living in the Light of Eternity

I

Eternity is, as a philosopher defines it, "an infinite extent of time, in which every event is future at one time, present at another, past at another." [1]

This is an interesting definition no doubt, but what is "infinity"? "No beginning and no end?" What is time that has no beginning and no end? Time cannot be defined without eternity nor eternity without time? Is eternity time going on forever in two directions, past-ward and future-ward? Is time eternity chopped to pieces or numbers?

Let us see whether a symbolic representation of eternity is more amenable to our understanding or imagination. What would a poet, for instance, say about it?

> I saw Eternity the other night,
> Like a great ring of pure and endless light,
>   All calm, as it was bright,
>   And round beneath it, Time, in hours, days, years
>   Driven by the spheres,
>   Like a vast shadow moved, in which the world
>   And all her train were hurled.[2]

Henry Vaughan's lines, as Bertrand Russell points

---

[1] *The Dictionary of Philosophy*, edited by Dagobert D. Runes (New York: Philosophical Library), p. 97.
[2] Henry Vaughan, "The World."

out,[3] are evidently suggested by Plato's *Timaeus* in which Plato states:

> Now the nature of the ideal being was everlasting, but to bestow this attribute in its fulness upon a creature was impossible. Wherefore he [God] resolved to have a moving image of eternity, and when he set in order the heaven, he made this image eternal but moving according to number, while eternity itself rests in unity; and this image we call time. For there were no days and nights and months and years before the heaven was created, but when he constructed the heaven he created them also.[4]

Further, Plato goes on to say that the heaven and time are so closely knit together that if one should dissolve the other might also be dissolved:

> Time, then, and the heaven came into being at the same instant in order that, having been created together, if ever there was to be a dissolution of them, they might be dissolved together. It was framed after the pattern of the eternal nature, that it might remember this as far as was possible; for the pattern exists from eternity, and the created heaven has been, and is, and will be, in all time.

The heaven is eternity; and "the sun and moon and the five stars" are "the forms of time, which imitate eternity and revolve according to a law of number," and the moving images of the eternal essence which alone "is" and not subject to becoming. What we see with our sense is not the heaven

---

[3] *History of Western Philosophy*, p. 144.

[4] *Dialogues of Plato*, translated by B. Jowett (London: Oxford University Press), Vol. III, p. 456. Published in the United States by Random House.

Living in the Light of Eternity

itself, the original eternal being itself, which is only in God's mind. If we wish, therefore, "to live in the light of eternity" we must get into God's mind. "Is this possible?" one may ask. But the question is not the possibility of achieving this end, but its necessity; for otherwise we cannot go on living even this life of ours though bound in time and measurable in days and nights, in months and years. What is necessary, then, must be possible. When the Eternal negated itself to manifest itself in "the forms of time," it assuredly did not leave the forms helpless all by themselves; it must have entered into them though negated. When the Eternal negated itself into the moving, changing, sensible forms of time, it hid itself in them. When we picked them up, we must see "the shoots of everlastingness" in them. "Was" and "will be" must be in "is." What is finite must be carrying in it, with it, everything belonging to infinity. We who are becoming in time, therefore, must be able to see that which eternally "is." This is seeing the world as God sees it, as Spinoza says, *"sub specie aeternitatis."*

Eternity may be regarded as a negation as far as human finitude is concerned, but inasmuch as this finitude is always changing, becoming, that is, negating itself, what is really negative is the world itself and not the eternal. The eternal must be an absolute affirmation which our limited human understanding defines in negative terms. We must see the world in this affirmation, which is God's way of seeing the world, seeing everything as part of the whole. "Living in the light of eternity" cannot be anything else.

B. Jowett, translator of Plato, writes in his introduction to *Timaeus*:

> Not only Buddhism, but Greek as well as Christian philosophy, show that it is quite possible that the human mind should retain an enthusiasm for mere negations. . . . Eternity or the eternal is not merely the unlimited in time but the truest of all Being, the most real of all realities, the most certain of all knowledge, which we nevertheless only see through a glass darkly.[5]

The enthusiasm Jowett here refers to is not "for mere negations" or for things which are "seen only through a glass darkly"; it cannot come out of the human side of finitude; it must issue from eternity itself, which is in the finitude, indeed, and which makes the finitude what it is. What appears to be a mere negation from the logical point of view is really the is-ness of things. As long as we cannot transcend the mere logicality of our thinking, there will be no enthusiasm of any kind whatever in any of us. What stirs us up to the very core of our being must come from the great fact of affirmation and not from negation.

2

Buddhism is generally considered negativistic by Western scholars. There is something in it which tends to justify this view, as we observe in Nāgārjuna's doctrine of "Eight No's":

> There is no birth,
> Nor is there death;
> There is no beginning,

[5] *Ibid.*, p. 398.

Nor is there any ending;
Nothing is identical with itself,
Nor is there any diversification;
Nothing comes into existence,
Nor does anything go out of existence.[6]

What he aims at by negating everything that can be predicated of the *Dharma* (Ultimate Reality) is to bring out thereby what he terms the Middle Way. The Middle Way is not sheer nothingness, it is a something that remains after every possible negation. Its other name is the Unattainable, and the *Prajñā-pāramitā-sūtra* teaches the doctrine of the Unattainable. I will try to illustrate what it means in order to clarify the deeper implications of this contradictory statement. I shall repeat the story found in Chapter II.

There was once in the T'ang dynasty in the history of China, a great scholar thoroughly versed in this doctrine. His name was Tokusan (790–865, Tê-shan in Chinese). He was not at all satisfied with the Zen form of Buddhist teaching which was rapidly gaining power, especially in the south of China. Wishing to refute it he came out of Szŭ-ch'uan in the southwestern part of China.

His objective was to visit a great Zen monastery in the district of Li-yang. When he approached it he thought of refreshing himself with a cup of tea. He entered a teahouse by the roadside and ordered some refreshments. Seeing a bundle on his back, the old lady who happened to be the teahouse keeper asked what it was.

Tokusan said, "This is Shoryo's [Ch'ing-lung's]

[6] The *Mādhyamika-śastra*, "Treatise on the Middle Way."

great commentary on the *Diamond Sūtra* [a portion of the great *Prajñā-pāramitā-sūtra*]."

"I have a question and if you answer it I shall be glad to serve you the refreshments free of charge. Otherwise, you will have to go elsewhere."

"What is your question?" the monk asked.

"According to the *Diamond Sūtra*, 'The past mind is unattainable, the future mind is unattainable, and the present mind is unattainable.' If so, what is the mind which you wish to punctuate?"

An explanation is needed here. In Chinese, "refreshments," *t'ien-hsin*, literally means "punctuating the mind." I do not know how the term originated. The teahouse keeper making use of "the mind" associated with "refreshments" quoted the *Sūtra* in which the mind in terms of time is said to be "unattainable" in any form, either past, present, or future. If this is the case, the monk cannot have any "mind" which he wishes to "punctuate." Hence her question.

Tokusan was nonplused, because he was never prepared to encounter such questions while studying the *Sūtra* along the conventional line of conceptual interpretation. He could not answer the question and was obliged to go without his tea. Those who do not know how to transcend time will naturally find it difficult to attain Nirvana which is eternity.

The unattainability of Nirvana comes from seeking it on the other shore of becoming as if it were something beyond time or birth-and-death (*samsāra*). Nirvana is *samsāra* and *samsāra* is Nirvana. Therefore, eternity, Nirvana, is to be grasped where time, *samsāra*, moves on. The refreshments cannot be taken outside time. The taking is time. The taking is something attainable, and yet it goes on in something unattainable. For without this something unattainable all that is attainable will cease to be attainable. This paradoxicality marks life.

Time is elusive, that is, unattainable. If we try to take hold of it by looking at it from the outside, then we cannot even have ordinary refreshments. When time is caught objectively in a serialism of past, present, and future, it is like trying to catch one's own shadow. This is negating eternity constantly. The unattainable must be grasped from the inside. One has to live in it and with it. While moving and changing, one must become the moving and changing. Emerson in "Brahma" sings of the eternal as "one" in the changing and moving forms of time:

> They reckon ill who leave me out;
>   When me they fly, I am the wings;
> I am the doubter and the doubt,
>   And I the hymn the Brahmin sings.

Where "the doubter and the doubt" are one, there is Brahma as "the pattern of the eternal nature," which is God himself. When "the doubter and the doubt" are separated and placed in the serialism of time, the dichotomy cuts into every moment of life darkening forever the light of eternity.

"Living in the light of eternity" is to get into the oneness and allness of things and to live with it. This is what the Japanese call "seeing things *sono-mama*" [7] in their suchness, which in William Blake's

---

[7] In the "as-it-is-ness" of things.

terms is to "hold infinity in the palm of your hand, and eternity is an hour."

To see things as God sees them, according to Spinoza, is to see them under the aspect of eternity. All human evaluation is, however, conditioned by time and relativity. It is ordinarily difficult for us humans "to see a world in a grain of sand, and a heaven in a wild flower." To our senses, a grain of sand is not the whole world, nor is a wild flower in a corner of the field a heaven. We live in a world of discrimination and our enthusiasm rises from the consideration of particulars. We fail to see them "evenly" or "uniformly" as Meister Eckhart tells us to do, which is also Spinoza's way, Blake's way, and other wise men's way, East and West. Tennyson must have been in a similar frame of consciousness when he plucked a wild flower out of the crannied wall and held it in his hand and contemplated it.[8]

3

However difficult this way of looking at the world is, the strange thing to most of us, or rather the wonderful thing, is that once in a while we transcend the temporal and relativistic point of view. It is then that we realize that life is worth living, and that death is not the end of all our strivings, and furthermore that what Buddhists call "thirst" (*tṛiṣṇā*) is more deeply rooted than we imagine, as it grows straight out of the root of *karuṇā*.[9]

Let me cite a Japanese Haiku poet of the eight-

[8] "Flower in the Crannied Wall."
[9] "Compassion." One may say it is the Buddhist equivalent of love.

Living in the Light of Eternity

eenth century, Bashō. One of his seventeen-syllable poems reads:

> When closely inspected,
> One notices a nazuna in bloom
> Under the hedge.

The nazuna is a small flowering wild plant. Even when flowering it is hardly noticeable, having no special beauty. But when the time comes, it blooms, fulfilling all that is needed of a living being as ordered at the beginning of creation. It comes directly from God as does any other form of being. There is nothing mean about it. Its humble glory surpasses all human artificiality. But ordinarily we pass by it and pay not the slightest attention. Bashō at the time must have been strangely impressed by it blooming under a thickly growing hedge, modestly lifting its tender head hardly discernible from the rest. The poet does not at all express his emotions. He makes no allusions whatever to "God and man," nor does he express his desire to understand "What you are root, and all, and all in all." He simply looks at the nazuna so insignificant and yet so full of heavenly splendor and goes on absorbed in the contemplation of "the mystery of being," standing in the midst of the light of eternity.

At this point it is important to note the difference between East and West. When Tennyson noticed the flower in a crannied wall he "plucked" it and held it in his hand and went on reflecting about it, pursuing his abstract thought about God and man, about the totality of things and the unfathomability of life. This is characteristic of Western man.

His mind works analytically. The direction of

his thinking is toward the externality or objectivity of things. Instead of leaving the flower as it is blooming in the cranny, Tennyson must pluck it out and hold it in his hand. If he were scientifically minded, he would surely bring it to the laboratory, dissect it, and look at it under the microscope; or he would dissolve it in a variety of chemical solutions and examine them in the tubes, perhaps over a burning fire. He would go through all these processes with anything, mineral or vegetable, animal or human. He would treat the human body, dead or alive, with the same innocence or indifference as he does a piece of stone. This is also a kind of seeing the world in the aspect of eternity or rather in the aspect of perfect "evenness."

When the scientist finishes (though the "when" of this is unpredictable) his examination, experimentation, and observation, he will indulge in all forms of abstract thinking; evolution, heredity, genetics, cosmogeny. If he is still more abstract-minded, he may extend his speculative mood to a metaphysical interpretation of existence. Tennyson does not go so far as this. He is a poet who deals with concrete images.

Compare all this with Bashō and we see how differently the Oriental poet handles his experience. Above all, he does not "pluck" the flower, he does not mutilate it, he leaves it where he has found it. He does not detach it from the totality of its surroundings, he contemplates it in its *sono-mama* state, not only in itself but in the situation as it finds itself—the situation in its broadest and deepest possible sense. Another Japanese poet refers to the wild flowers:

> All these wild flowers of the fields—
> Should I dare touch them?
> I offer them as they are
> To all the Buddhas in the
> Three thousand chiliocosms!

Here is the feeling of reverence, of mystery, of wonderment, which is highly religious. But all this is not expressly given articulation. Bashō simply refers first to his "close inspection" which is not necessarily aroused by any purposeful direction of his intention to find something among the bushes; he simply looks casually around and is greeted unexpectedly by the modestly blooming plant which ordinarily escapes one's detection. He bends down and "closely" inspects it to be assured that it is a nazuna. He is deeply touched by its unadorned simplicity, yet partaking in the glory of its unknown source. He does not say a word about his inner feeling, every syllable is objective except the last two syllables, "*kana*." [10] "*Kana*" is untranslatable into English, perhaps except by an exclamation mark, which is the only sign betraying the poet's subjectivity. Of course, a Haiku being no more than a poem of seventeen syllables cannot express everything that went on in Bashō's mind at the time. But this very fact of the Haiku's being so extremely epigrammatic and sparing of words gives every syllable used an intensity of unexpressed inner feeling of the poet, though much is also left to the reader to discover what is hidden between the sylla-

---

[10] *Yoku mireba*   [When] carefully seen,
    Nazuna hana saku   Nazuna in bloom,
    Kakine kana!   The hedge!

bles. The poet alludes to a few significant points of reference in his seventeen-syllable lines leaving the inner connection between those points to be filled by the sympathetically or rather empathically vibrating imagination of the reader.

4

Western psychologists talk about the theory of empathy or transference of feeling or participation, but I am rather inclined to propound the doctrine of identity. Transference or participation is based upon the dualistic interpretation of reality whereas the identity goes more fundamentally into the root of existence where no dichotomy in any sense has yet taken place. From this point of view, participation becomes easier to understand and may be more reasonable or logical. For no participation is possible where there is no underlying sense of identity. When difference is spoken of, this presupposes oneness. The idea of two is based on that of one. Two will never be understood without one. To visualize this, read the following from Traherne's *Centuries of Meditations*:

> You never enjoy the world aright, till the Sea itself floweth in your veins, till you are clothed with the heavens, and crowned with the stars: and perceive yourself to be the sole heir of the whole world, and more than so, because men are in it who are every one sole heirs as well as you.[11]

---

[11] *Centuries of Meditations*, Thomas Traherne, 1636–1674 (London: P. J. & A. E. Dobell), p. 19.

Living in the Light of Eternity 117

Or this:

> Your enjoyment of the world is never right, till every morning you awake in Heaven; see yourself in your Father's Palace; and look upon the skies, the earth, and the air as Celestial Joys; having such a reverend esteem of all, as if you were among the Angels.[12]

Such feelings as these can never be comprehended so long as the sense of opposites is dominating your consciousness. The idea of participation or empathy is an intellectual interpretation of the primary experience, while as far as the experience itself is concerned, there is no room for any sort of dichotomy. The intellect, however, obtrudes itself and breaks up the experience in order to make it amenable to intellectual treatment, which means a discrimination or bifurcation. The original feeling of identity is then lost and intellect is allowed to have its characteristic way of breaking up reality into pieces. Participation or empathy is the result of intellectualization. The philosopher who has no original experience is apt to indulge in it.

According to John Hayward, who wrote an introduction to the 1950 edition of Thomas Traherne's *Centuries of Meditations*, Traherne is "a theosopher or visionary whose powerful imagination enabled him to see through the veil of appearances and rediscover the world in its original state of innocence." This is to revisit the Garden of Eden, to regain Paradise, where the tree of knowledge has not yet begun to bear fruit. The Wordsworthian "Intimations" are no more than our longings for eternity that was left behind. It is our

[12] *Ibid.*

eating the forbidden fruit of knowledge which has resulted in our constant habit of intellectualizing. But we have never forgotten, mythologically speaking, the original abode of innocence; that is to say, even when we are given over to intellection and to the abstract way of thinking, we are always conscious, however dimly, of something left behind and not appearing on the chart of well-schematized analysis. This "something" is no other than the primary experience of reality in its suchness or is-ness, or in its *sono-mama* state of existence. "Innocence" is a Biblical term and corresponds ontologically to "being *sono-mama*" as the term is used in Buddhism.

Let me quote further from Traherne whose eternity-piercing eye seems to survey the beginningless past as well as the endless future. His book of "meditations" is filled with wonderful insights born of a profound religious experience which is that of one who has discovered his primal innocence.

> Will you see the infancy of this sublime and celestial greatness? Those pure and virgin apprehensions I had from the womb, and that divine light wherewith I was born are the best unto this day, wherein I can see the Universe. . . .
>
> Certainly Adam in Paradise had not more sweet and curious apprehensions of the world, than I when I was a child.
>
> My very ignorance was advantageous. I seemed as one brought into the Estate of Innocence. All things were spotless and pure and glorious: yea, and infinitely mine, and joyful and precious. I knew not that there were any sins, or complaints or laws. I dreamed not of poverties,

contentions or vices. All tears and quarrels were hidden from mine eyes. Everything was at rest, free and immortal, I knew nothing of sickness or death or rents or exaction, either tribute or bread. . . .

All Time was Eternity, and a perpetual Sabbath. . . .

All things abided eternally as they were in their proper places. Eternity was manifest in the Light of the Day, and something infinite behind everything appeared: which talked with my expectation and moved my desire. The city seemed to stand in Eden, or to be built in Heaven. . . .

5

Compared with these passages, how prosaic and emotionally indifferent Zen is! When it sees a mountain it declares it to be a mountain; when it comes to a river, it just tells us it is a river. When Chokei (Chang-ching) after twenty hard years of study happened to lift the curtain and saw the outside world, he lost all his previous understanding of Zen and simply made this announcement:

> How mistaken I was! How mistaken I was!
> Raise the screen and see the world!
> If anybody asks me what philosophy I understand,
> I'll straightway give him a blow across his mouth
>   with my *hossu*.

Chokei does not say what he saw when the screen was lifted up. He simply resents any question being asked about it. He even goes to the length of keeping the questioner's mouth tightly closed. He knows that if one even tried to utter a word and say "this" or "that," the very designation misses the mark. It is like another master's bringing out before the entire congregation a monk who asked him who Buddha was. The master then made this remark, "Where does this monk want to find Buddha? Is this not a silly question?" Indeed, we are all apt to forget that every one of us is Buddha himself. In the Christian way of saying, this means that we are all made in the likeness of God or, in Eckhart's words, that "God's is-ness is my is-ness and neither more nor less." [13]

It may not be altogether unprofitable in this connection to give another Zen "case" where God's is-ness is made perceivable in the world of particulars as well as in the world of absolute oneness. To us the case illustrates the Eckhartian knowledge "that I know God as He knows me, neither more nor less but always the same." This is knowing things as they are, loving them in their *sono-mama* state, or "loving justice for its own sake," [14] that is to say, "loving God without any reason for loving." Zen may look so remote and aloof from human affairs that between it and Eckhart some may be persuaded to see nothing of close relationship as I

[13] Blakney, p. 180.
[14] Eckhart's idea of "justice" may be gleaned from the following passages from his "Sermon" 18 (Blakney, pp. 178–82):

"He is just who gives to each what belongs to him."

"They are just who take everything from God evenly, just as it comes, great and small, desirable and undesirable, one thing like another, all the same, and neither more nor less."

"The just live eternally with God, on a par with God, neither deeper nor higher."

"God and I: we are one. By knowing God I take him to myself. By loving God I penetrate him."

Living in the Light of Eternity

am trying to show here. But in reality Eckhart uses in most cases psychological and personalistic terms whereas Zen is steeped in metaphysics and in transcendentalism. But wherever the identity of God and man is recognized the Zen statements as they are given below will be intelligible enough.

Hakuin (1685–1768), a great Japanese Zen master of the Tokugawa era, quotes in his famous book known as *Kwai-an-koku Go* (fas. 5) a story of Shun Rofu's interview with a well-seasoned lay disciple of Zen. Shun (of the Sung dynasty) was still a young man when this interview took place. It was the custom of this lay disciple to ask a question of a new monk-visitor who wanted to enjoy the hospitality of the devoted Zen Buddhist, and the following once took place between him and a new caller:

Q. "How about the ancient mirror which has gone through a process of thorough polishing?"

A. "Heaven and earth are illuminated."

Q. "How about before the polishing?"

A. "As dark as black lacquer."

The layman Buddhist was sorry to dismiss the monk as not fully deserving his hospitality.

The monk now returned to his old master and asked:

Q. "How about the ancient mirror not yet polished?"

A. "Han-yang is not very far from here."

Q. "How about after the polish?"

A. "The Isle of Parrot [Ying-wu] lies before the Pavilion of Yellow Stork [Huang-huo]."

This is said to have at once opened the monk's eye to the meaning of the ancient mirror, which was the subject of discussion between him and Shun. "The mirror" in its is-ness knows no polishing. It is the same old mirror whether or not it goes through any form of polishing. "Justice is even," says Eckhart. For "the just have no will at all: whatever God wants, it is all one to them."

Now Hakuin introduces the following *mondo:*[15]

A monk asked Ho-un of Rosozan, a disciple of Nangaku Yejo (died 744), "How do we speak and not speak?" This is the same as asking: How do we transcend the law of contradiction? When the fundamental principle of thought is withheld, there will be no thinking of God as Eckhart tells us, "God [who] is in his own creature—not as he is conceived by anyone to be—not yet as something yet to be achieved—but more as an 'is-ness,' as God really is."[16] What kind of God can this be? Evidently, God transcends all our thought. If so, how have we ever come to conceive of God? To say God is "this" or "that" is to deny God, according to Eckhart. He is above all predicates, either positive or negative. The monk's question here ultimately brings us to the same form of quandary.

Ho-un of Rosozan, instead of directly answering the monk, retorted, "Where is your mouth?"

The monk answered, "I have no mouth." Poor monk! He was aggressive enough in his first questioning, for he definitely demanded to get an answer to the puzzle: "How could reality be at once

[15] "Question and answer."
[16] Blakney, p. 204.

Living in the Light of Eternity

an affirmation and a negation?" But when Ho-un counterquestioned him, "Where is your mouth?" all that the monk could say was, "I have no mouth." Ho-un was an old hand. Detecting at once where the monk was, that is, seeing that the monk was still unable to transcend the dichotomy, Ho-un pursued with "How do you eat your rice?"

The monk had no response. (The point is whether he had a real understanding of the whole situation.)

Later Tozan, another master, hearing of this *mondo*, gave his own answer: "He feels no hunger and has no need for rice."

"One who feels no hunger" is "the ancient mirror" that needs no polishing, is he who "speaks and yet speaks not." He is "justice" itself, the justice is the suchness of things. To be "just" means to be *sono-mama*, to follow the path of "everyday consciousness," "to eat when hungry and to rest when tired." In this spirit I interpret Eckhart's passage: "If I were perpetually doing God's will, then I would be a virgin in reality, as exempt from idea-handicaps as I was before I was born." [17] "Virginity" consists in not being burdened with any forms of intellection, in responding with "Yes, yes" when I am addressed by name. I meet a friend in the street, he says, "Good morning," and I respond, "Good morning." This will again correspond to the Christian way of thinking: "If God told an angel to go to a tree and pick off the caterpillar, the angel would be glad to do it and it would be bliss to him because it is God's will." [18]

[17] *Ibid.*, p. 207.
[18] *Ibid.*, p. 205.

A monk asked a Zen master, "I note an ancient wise man saying: 'I raise the screen and face the broad daylight; I move the chair and am greeted by the blue mountain.' What is meant by 'I raise the screen and face the broad daylight'?"

The master said, "Please pass me the pitcher there."

"What is meant by 'I move the chair and am greeted by the blue mountains'?"

"Please put the pitcher back where it was found." This was the answer given by the master.

All these Zen *mondo* may sound nonsensical and the reader may come to the conclusion that when the subject is "living in the light of eternity" they are altogether irrelevant and have no place in a volume like this. It is quite a natural criticism from the point of view of an ordinary man of the world. But let us listen to what Eckhart, one of the greatest mystics in the Christian world, states about the "now-moment" which is no other than eternity itself:

The now-moment in which God made the first man, and the now-moment in which the last man will disappear, and the now-moment in which I am speaking are all one in God, in whom there is only one now.[19]

I have been reading all day, confined to my room, and feel tired. I raise the screen and face the broad daylight. I move the chair on the veranda and look at the blue mountains. I draw a long breath, fill my lungs with fresh air and feel entirely refreshed.

---

[19] *Ibid.*, p. 209.

Living in the Light of Eternity

I make tea and drink a cup or two of it. Who would say that I am not living in the light of eternity? We must, however, remember that all these are events of one's inner life as it comes in touch with eternity or as it is awakened to the meaning of "the now-moment" which is eternity, and further that things or events making up one's outer life are no problems here.

6

I quote again from Eckhart's Sermon 18:

In eternity, the Father begets the Son in his own likeness. "The Word was with God and the Word was God." Like God, it had his nature. Furthermore, I say that God has begotten him in my soul. Not only is the soul like him and he like it, but he is in it, for the Father begets his Son in the soul exactly as he does in eternity and not otherwise. He must do so whether he will or not. The Father ceaselessly begets his Son and, what is more, he begets me not only as his Son but as himself and himself as myself, begetting me in his own nature, his own being. At that inmost Source, I spring from the Holy Spirit and there is one life, one being, one action. All God's works are one and therefore He begets me as he does his Son and without distinction.[20]

Is this not a strong bold saying? But there is no denying its absolute truth. Yet we must not forget that the truth of Eckhart's sermon comes from setting ourselves in the light of eternity. As long as we are creatures in time and seeking our own and not God's will, we shall never find God in ourselves.

[20] *Ibid.*, p. 181.

When references are made to Christian symbolism such as "God," "Father," "Son," "Holy Spirit," "begetting," and "likeness," the reader may wonder in what sense Buddhists are using these terms. But the truth is that symbols are after all symbols and when this inner signification is grasped they can be utilized in any way one may choose. First, we must see into the meaning and discard all the historical or existential encumbrances attached to the symbols and then we all, Christians as well as Buddhists, will be able to penetrate the veil.

The Biblical God is said to have given his name to Moses on Mount Sinai as "I am that I am." This is a most profound utterance, for all our religious or spiritual or metaphysical experiences start from it. This is the same as Christ's saying, "I am," that is, he is eternity itself, while Abraham is in time, therefore, he "was" and not "is." Those who live in the light of eternity always are and are never subjected to the becoming of "was" and "will be."

Eternity is the absolute present and the absolute present is living a *sono-mama* life, where life asserts itself in all its fullness.

Appendices

chapter 5

Transmigration

Does Buddhism teach transmigration? If it does, how does it work? Does the soul really transmigrate?

Such questions are frequently asked, and I will try briefly to answer them here.

I

The idea of transmigration is this: After death, the soul migrates from one body to another, celestial, human, animal, or vegetative.

In Buddhism, as it is popularly understood, what regulates transmigration is ethical retribution. Those who behave properly go to heaven, or to heavens, as there are many heavens according to Buddhist cosmology. Some may be reborn among their own races. Those, however, who have not conducted themselves according to moral precepts will be consigned after death to the underground worlds called Naraka.

There are some destined to be reborn as a dog or a cat or a hog or a cow or some other animal, according to deeds which can be characterized as preeminently in correspondence with those natures generally ascribed to those particular animals. For instance, the hog is popularly thought to be greedy and filthy. Thus those of us who are especially inclined to be that way will be hogs in their next lives. Others who are rather smart or cunning or somewhat mischievous may be born as rats or monkeys or foxes. This reminds us of Swedenborg's doctrine of correspondence, according to which things on earth have corresponding things in heaven or hell.

Sometimes we are said to be born as plants or even rocks.

The interesting thing about this idea of transmigration as sometimes told by Buddhists is that we do not stay in heaven or hell forever. When our karma is exhausted, we come out of hell or come down from heaven. Even when we turn into cats or dogs, we do not repeat this kind of life all the time. We may be reborn as human beings again if we do something good while living as a lower animal, though it is highly doubtful that, for instance, the cat can be taught not to steal fish from the neighbors—which is what she does quite frequently in Japan—however well she may be fed at home.

But so far nobody has advanced the method of calculating mathematically the strength of karma according to the character of each deed. Therefore, we can never tell how long our life in heaven or hell will be. In any case, we know this much: there is a time when we have to leave heaven or hell.

Buddhists are more concerned—which is natural—with Naraka (hells) than heavens. After death we generally go to Yama, who rules the spirits of the dead. He is known as Emma-sama in Japanese. He has a bright mirror before him. When we appear before him, we see ourselves reflected in it. It illuminates our entire being, and we cannot hide anything from it. Good and bad, all is reflected in it as it is. Emma-sama looks at it and knows at once what kind of person each of us was while living in the world. Besides this, he has a book before him in which everything we did is minutely recorded. We are therefore before the Lord of Death exactly what we were, and there is no deceiving him. His judgment goes straight to the core of our personality. It never errs. His penetrating eye reads not only our consciousness but also our unconscious. He is naturally legalistic, but he is not devoid of kindheartedness, for he is always ready to discover in the unconscious something which may help the criminal to save himself.

2

The idea of transmigration has a certain appeal to the imaginative mind if one is not too critical or scientific—the idea that each motive, consciously conceived or unconsciously prompted, has its ethical value and is punishable or rewardable accordingly, and that the Lord of Death ruling the underworld makes no mistake in assigning us to places where we each belong. His mirror of judgment and his records never err in this respect. These ideas correspond to our sense of justice and compensation. Instead of all sinners being summarily consigned to everlasting fire when the Day of Judgment comes, it is certainly more in accord with common sense and justice that each sin, judiciously weighed and evaluated, be given its particular due. This evaluation and consignment, when demonstrated in the doctrine of transmigration, takes on a poetic coloring.

Suppose I did something wrong or something not so very bad and were made to be reborn as a cat. I would live in this animal form for a while, perhaps eight or ten years, for the cat does not live very long. My sin is expiated, for probably I behaved properly as a cat from the human point of view. As a reward, I am born again as a human being. Now, if I remembered this experience as a cat, would it not be highly interesting for me as a former cat to observe all that the mother cat now in my house does, playing with her kittens, sometimes bringing a lizard and even a little snake from the yard for the little ones to play with?

When not only the cat but all the other animals, and also plants and rocks, are looked upon from this point of view, that is, as possible forms of our reincarnation in the future as well as in the past, would not our interest in all those objects existing about us take quite a new turn and perhaps become a source of spiritual inspiration in some way?

For one thing, those forms surrounding us cease to be things altogether foreign to us. They are not strangers; they are not something hostile. On the contrary, they share our nature. We are ready to transform ourselves into their forms of existence, and they too can someday take human form when they are so conditioned. There is a mutual interest between us and them. There is a bond of sympathy and mutual understanding between human beings and the rest of the world.

Besides these considerations, the doctrine of transmigration affords us the chance of pilgrimaging throughout the whole universe, from the thirty-three heavens to the nineteen hells, including the other realms such as the *tiryagyona* (animal), *preta* (hungry ghost), and *asura* (fighting devils). While it is not at all pleasant to be fighting all the time, to be tortured in various ways, or to be eternally hungry, it is in accord with human nature to experience vicissitudes of existence and thereby to learn to read the meaning of life.

Nobody likes to be in hell and tortured. But because of this experience, we know how to appreciate heavenly pleasures and how to be sympathetic with our fellow beings who happen to be in not so pleasant an environment.

3

Transmigration pictures us traveling through an infinite number of Kalpas as we go individually experiencing life in its possible varieties. Evolution, however, delineates human existence as a whole as having gone through all these stages. This is the difference between science and religion: science deals with abstractions, whereas religion is individualistic and personal. So far, evolution has not taken account of ethical implications. It has treated the subject from the point of view of biology and psychology. In the rising development of the human race, the scientists have not given much significance to the ethical and spiritual factors; they have been primarily concerned with the way man has made use of his intelligence more than anything else in his so-called upward course of development.

Transmigration reviews man's existence entirely

from the point of view of ethics and religion; it is hardly concerned with his intelligence. And this is the very point where transmigration interests us. The idea may not have anything deserving scientific investigation. But in spite of this, it perpetually attracts the attention of religious minded people.

4

Theoretically speaking, the idea of incarnation must have come first, then reincarnation, and finally transmigration. Something took the flesh, God or the word or the devil or the first principle or anything else, which had to express itself in a tangible and visible form so that we can talk of it as something. Being made of the senses and intellect, we individualize, which means incarnation.

When incarnation is established, reincarnation is easy to follow; and when reincarnation is morally evaluated, we have transmigration. Transmigration then comes to be connected with the idea of punishment and reward.

There is another implication of transmigration, which is the idea of the moral perfectibility of human nature. Before Buddha attained Buddhahood he went through many an incarnation, and in each reincarnation he is said to have practiced the six or ten virtues of *pāramitā*, whereby in his last incarnation as a human being he became a perfect man, that is, Buddha.

As long as we have the idea of an infinite possibility of perfecting ourselves morally, we must find some way of carrying this idea through. Inas-

Transmigration

much as we cannot forever continue our individual existence as such, there must be another way of solving the problem, which is what we may call the eternally progressive conception of transmigration.

5

Besides this interpretation of the transmigration idea in its moral and punitive aspects, there is an enjoyable phase of it when we make it a matter of experience during our lifetime. When we scrutinize our daily experiences, we realize that we have here everything we could experience by going through an indefinitely long period of transmigration. Every shade of feeling we have while on earth finds its counterpart somewhere in the heavens or in the hells or in some intermediate realms of the *preta*, or *asura* or *tiryagyona*. For instance, when we are angry, we are with the *asura;* when we are pleased, we are transported into the heaven of joy, *nirmanarataya;* when we are restless, we have turned into the monkey; when we can imagine ourselves free from guilt, we bloom as the lotus or as the morning glory in the early summer dawn, and so on. The whole universe depicts itself in human consciousness. That is to say, our daily life is an epitome of an indefinitely long career of transmigration.

6

As far as I can see, the doctrine of transmigration does not seem to enjoy any scientific support. The first question we encounter is, "What is it that transmigrates?" We may answer, "It is the soul." "What, then, is the soul?" The soul cannot be conceived as an entity or an object like any other objects we see about us. It cannot be anything tangible or visible. If so, how does it manage to enter into a body? How does it get out of one body when this body decomposes and pass into another body? Where is this "other body" waiting for the liberated soul to enter? The body without the soul is inconceivable; we cannot imagine a soulless body in existence somewhere to receive the soul newly detached. If the soul can maintain itself without embodying itself, why do we not find bodyless souls wandering somewhere? Can a soul subsist without a body?

If the doctrine of transmigration is to be tenable, we must say that there is something that transmigrates; if there is something, what is it? If we cannot affirm it as an entity, what can it be? Can the questions enumerated above be satisfactorily answered? There are still other questions which must be answered before we can establish transmigration.

7

We can think of the soul not as an entity but as a principle. We can conceive of the soul as not entering into a body already in existence and ready to receive the soul, but as creating a body suitable for its own habitation. Instead of form or structure determining function, we can take function as determining form. In this case, the soul comes first and the body is constructed by it. This is really the Buddhist conception of transmigration.

Buddhist philosophy considers *tṛiṣṇā* or *taṇhā*, or "thirst," the first principle of making things come into existence. In the beginning there is *tṛiṣṇā*. It wills to have a form in order to express itself, which means to assert itself. In other words, when it asserts itself it takes form. As *tṛiṣṇā* is inexhaustible, the forms it takes are infinitely varied. *Tṛiṣṇā* wants to see and we have eyes; it wants to hear and we have ears; it wants to jump and we have the deer, the rabbit, and other animals of this order; it wants to fly and we have birds of all kinds; it wants to swim and we have fish wherever there are waters; it wants to bloom and we have flowers; it wants to shine and we have stars; it wants to have a realm of heavenly bodies and we have astronomy; and so on. *Tṛiṣṇā* is the creator of the universe.

Being the creator, *tṛiṣṇā* is the principle of individuation. It creates a world of infinite diversities. It will never exhaust itself. We as its highest and richest expression can have an insight into the nature of *tṛiṣṇā* and its working. When we really see into ourselves, *tṛiṣṇā* will bare itself before itself in us. As it is not an individualized object, self-inspection is the only way to approach it and make it reveal all its secrets. And when we know them, perhaps we may also understand what transmigration really means.

When we see the lilies of the field and observe that they are more gloriously arrayed than Solomon in his day, is this not because in our *tṛiṣṇā* there is something participating in the *tṛiṣṇā* of the flower? Otherwise, we could never appreciate them. When we follow the fowls of the air and think of their being utterly free from care or worry, is this not because the pulse of our *tṛiṣṇā* beats in unison with the *tṛiṣṇā* of the fowls? If this were not the case, how could we ever come to the understanding of those creatures? Even when Nature is regarded as hostile, there must be something in it which calls out this feeling in us—which is to say, Nature partakes of (human) *tṛiṣṇā*.

The atom may be considered nothing but a cluster of electrically charged particles and having nothing in common with human *tṛiṣṇā*. But does it not respond to the appliances contrived by human minds and human hands? And is it not because of this response that we can read into the nature of the atom and even devise a weapon most destructive to us human beings? The atom certainly has its *tṛiṣṇā*, and it is this *tṛiṣṇā* that enables man to express it in a mathematical formula.

8

When I was discussing this subject the other day, one of the great thinkers now in America remarked, "Does this mean that there are in our consciousness all these *tṛiṣṇā* as its constituent elements?" This is perhaps the way most of our readers would like to interpret my presentation of *tṛiṣṇā* when I make it the basis of mutual understanding, as it were, between ourselves and Nature generally. But I must say that that is not the way I conceive *tṛiṣṇā*. *Tṛiṣṇā* lies in us not as one of the factors constituting our consciousness, but it is our being itself. It is I; it is you; it is the cat; it is the tree; it is the rock; it is the snow; it is the atom.

9

Some may like to compare *tṛiṣṇā* with Schopenhauer's Will to live, but my idea of *tṛiṣṇā* is deeper than his Will. For the Will as he conceives it is already differentiated as the Will striving to live against death, against destruction. The Will implies a dualism. But *tṛiṣṇā* remains still dormant, as it were, as in the mind of God, for God has not yet moved to his work of creation. This moving is *tṛiṣṇā*. It is *tṛiṣṇā* that moves. It is *tṛiṣṇā* that made God give out his fiat, "Let there be light." *Tṛiṣṇā* is what lies at the back of Schopenhauer's Will. *Tṛiṣṇā* is a more fundamental conception than the Will.

For Schopenhauer, the Will is blind; but *tṛiṣṇā* is neither blind nor not blind, for neither of them can yet be predicated of *tṛiṣṇā*. *Tṛiṣṇā* is not yet a what. It can be called the pure will. In early Buddhism, *tṛiṣṇā* forms one of the links in the chain of "Dependent Origination," and it is demanded of us to get rid of it in order that we may be freed from grief and fear. But early Buddhists were not logical enough to push the idea of *tṛiṣṇā* far enough to its very source. Their effort to deliver themselves from *tṛiṣṇā*'s so-called leading to grief, fear, and so on, was also the working of *tṛiṣṇā* itself. As long as we are human beings, we can never do away with *tṛiṣṇā*, or, as they say, destroy it. The destruction of *tṛiṣṇā* will surely mean the annihilation of ourselves, leaving no one who will be the enjoyer of the outcome. *Tṛiṣṇā* is indeed the basis of all existence. *Tṛiṣṇā* is existence. *Tṛiṣṇā* is even before existence.

Later Buddhists realized this truth and made *tṛiṣṇā* the foundation of their new system of teaching with its doctrines of the Bodhisattva, universal salvation, Amitābha's "vow" (*praṇidhāna*), the *pariṇāmanā* ("turning over of merit"), and so on. These are all the outgrowth of *tṛiṣṇā*. When a Zen master was asked, "How could one get away with *tṛiṣṇā*?" he answered, "What is the use of getting away with it?" He further said, "Buddha is Buddha because of it," or, "Buddha is *tṛiṣṇā*." In fact, the whole life of Śākyamuni illustrates this.

10

Coming back to the transmigration phase of the *tṛiṣṇā* doctrine, I should like to assert again that this *tṛiṣṇā* as it expresses itself is essentially the same in any form it may take. (We cannot think of it in any other way.) The human *tṛiṣṇā* as we feel it inwardly must be that of the cat, or the dog, or the crow, or the snake. When a cat runs after a rat, when a snake devours a frog, when a dog jumps up furiously barking at a squirrel in the tree, when a pig goes around groveling in the mud, when the fish swims about contentedly in the pond, when the waves rage angrily on a stormy ocean, do we not feel here our own *tṛiṣṇā* expressing some of its infinitely variable modes? The stars are shining brightly, wistfully twinkling in a clear autumnal night; the lotus flowers bloom in the early summer morning even before the sun rises; when the spring comes, all the dead trees vie with one another to shoot out their fresh green leaves, waking up from a long winter sleep—do we not see here also some of our human *tṛiṣṇā* asserting itself?

I do not know whether ultimate reality is one or two or three or many more, but I feel that one *tṛiṣṇā*, infinitely diversified and diversifiable, expresses itself making up this world of ours. As *tṛiṣṇā* is subject to infinite diversifications, it can take infinitely variable forms. It is *tṛiṣṇā*, therefore, that determines form and structure. This is what is given to our consciousness, and our consciousness is the last word, we cannot go any further.

Viewing the idea of transmigration from this standpoint, is it not interesting to realize that we are practicing this transmigration in every moment of our lives, instead of going through it after death and waiting for many a Kalpa to elapse?

I do not know whether transmigration can be proved or maintained on the scientific level, but I know that it is an inspiring theory and full of poetic suggestions, and I am satisfied with this interpretation and do not seem to have any desire to go beyond it. To me, the idea of transmigration has a personal appeal, and as to its scientific and philosophical implications, I leave it to the study of the reader.

11

It may not be amiss here to add a word regarding the difference of attitude between the earlier and the later Buddhists toward the doctrine of transmigration and *tṛiṣṇā*. As we have already seen, the earlier Buddhist treatment of the subject is always negative, for it tends to emphasize the aspect of liberation or emancipation. The later Buddhists, however, have turned against this and strongly insist on *tṛiṣṇā* as being most fundamental and primary and needed for the general welfare not only of mankind but of all other beings making up the entire world. They would declare that *tṛiṣṇā* works in the wrong way when it chooses bad associates; that is, when it combines itself with the relative or psychological self, relying on the latter as the ultimate reality and as the controlling principle of life. *Tṛiṣṇā* then turns into the most ungovernable and insatiable upholder of power. What the earlier Buddhists wanted to conquer was this kind of *tṛiṣṇā*, swerved from its primal nature and becoming the thrall of egotistic impulses. Indeed, they wished, instead of conquering it, to escape from this state of thralldom. This made them negativists and escapists.

The later Buddhists realized that *tṛiṣṇā* was what constituted human nature—in fact, everything and anything that at all comes into existence—and that to deny *tṛiṣṇā* was committing suicide; to escape from *tṛiṣṇā* was the height of contradiction or a deed of absolute impossibility; and that the very thing that makes us wish to deny or to escape from *tṛiṣṇā* was *tṛiṣṇā* itself. Therefore, all that we could do for ourselves, or rather all that *tṛiṣṇā* could do for itself, was to make it turn to itself, to purify itself from all its encumbrances and defilements, by means of transcendental knowledge (*prajñā*). The later Buddhists then let *tṛiṣṇā* work on in its own way without being impeded by anything else. *Tṛiṣṇā* or "thirst" or "craving" then comes to be known as *mahākaruṇā*, or "absolute compassion," which they consider the essence of Buddhahood and Bodhisattvahood.

This *tṛiṣṇā* emancipated from all its encumbrances incarnates itself in every possible form in order to achieve a universal salvation of all beings, both sentient and non-sentient. Therefore, when Buddha declares that he is "all-conquering, all-knowing" he means that he has *tṛiṣṇā* in its purity. For when *tṛiṣṇā* comes back to itself, it is all-conqueror and all-knower, and also all-loving. It is this love or *karuṇā* or *maitrī* that makes the Buddha or Bodhisattva abandon his eternally entering into a state of emptiness (*śūnyatā*) and subjects himself to transmigrate through the triple world. But in this case it is better not to call it transmigration but incarnation. For he assumes all kinds of form on his own account, that is, voluntarily, in order to achieve a universal salvation. He is then not any more a passive sufferer of karmic causation. He is the "tent-designer" (*gahakāraka*) himself.[1]

---

[1] *The Dhammapada*, verses 151-2.

chapter 6

Crucifixion and Enlightenment

I

Whenever I see a crucified figure of Christ, I cannot help thinking of the gap that lies deep between Christianity and Buddhism. This gap is symbolic of the psychological division separating the East from the West.

The individual ego asserts itself strongly in the West. In the East, there is no ego. The ego is nonexistent and, therefore, there is no ego to be crucified.

We can distinguish two phases of the ego-idea. The first is relative, psychological, or empirical. The second is the transcendental ego.

The empirical ego is limited. It has no existence of its own. Whatever assertion it makes, it has no absolute value; it is dependent on others. This is no more than the relative ego and a psychologically established one. It is a hypothetical one; it is subject to all kinds of conditions. It has, therefore, no freedom.

What is it, then, that makes it feel free as if it were really so independent and authentic? Whence this delusion?

The delusion comes from the transcendental ego being mistakenly viewed as it works through the empirical ego and abides in it. Why does the transcendental ego, thus mistakenly viewed, suffer itself to be taken for the relative ego?

The fact is that the relative ego which corresponds to the *manovijñāna* of the Yogacara school has two aspects of relationship, outer and inner.

Objectively speaking, the empirical or relative ego is one of many other such egos. It is in the world of plurality; its contact with others is intermittent, mediated, and processional. Inwardly, its contact or relationship with the transcendental ego is constant, immediate, and total. Because of this the inner relationship is not so distinctly cognizable as the outer one—which, however, does not mean that the cognition is altogether obscure and negligible and of no practical worth in our daily life.

On the contrary, the cognition of the transcendental ego at the back of the relative ego sheds light into the source of consciousness. It brings us in direct contact with the unconscious.

It is evident that this inner cognition is not the ordinary kind of knowledge which we generally have about an external thing.

The difference manifests itself in two ways. The object of ordinary knowledge is regarded as posited in space and time and subject to all kinds of scientific measurements. The object of the inner cognition is not an individual object. The transcendental ego cannot be singled out for the relative ego to be inspected by it. It is so constantly and immediately contacted by the relative ego that when it is detached from the relative ego it ceases to be itself. The transcendental ego is the relative ego and the relative ego is the transcendental ego; and yet they

Crucifixion and Enlightenment

are not one but two; they are two and yet not two. They are separable intellectually but not in fact. We cannot make one stand as seer and the other as the seen, for the seer is the seen, and the seen is the seer.

When this unique relationship between the transcendental ego and the relative ego is not adequately comprehended or intuited, there is a delusion. The relative ego imagines itself to be a free agent, complete in itself, and tries to act accordingly.

The relative ego by itself has no existence independent of the transcendental ego. The relative ego is nothing. It is when the relative ego is deluded as to its real nature that it assumes itself and usurps the position of the one behind it.

It is true that the transcendental ego requires the relative ego to give itself a form through which the transcendental ego functions. But the transcendental ego is not to be identified with the relative ego to the extent that the disappearance of the relative ego means also the disappearance of the transcendental ego. The transcendental ego is the creative agent and the relative ego is the created. The relative ego is not something that is prior to the transcendental ego standing in opposition to the latter. The relative ego comes out of the transcendental ego and is wholly and dependently related to the transcendental ego. Without the transcendental ego, the relative ego is zero. The transcendental ego is, after all, the mother of all things.

The Oriental mind refers all things to the transcendental ego, though not always consciously and analytically, and sees them finally reduced to it, whereas the West attaches itself to the relative ego and starts from it.

Instead of relating the relative ego to the transcendental ego and making the latter its starting point, the Western mind tenaciously clings to it. But since the relative ego is by nature defective, it is always found unsatisfactory and frustrating and leading to a series of disasters, and as the Western mind believes in the reality of this troublemaker, it wants to make short work of it. Here we can also see something characteristically Western, for they have crucified it.

In a way the Oriental mind is not inclined toward the corporeality of things. The relative ego is quietly and without much fuss absorbed into the body of the transcendental ego. That is why we see the Buddha lie serenely in Nirvana under the twin Sala trees, mourned not only by his disciples but by all beings, non-human as well as human, non-sentient as well as sentient. As there is from the first no ego-substance, there is no need for crucifixion.

In Christianity crucifixion is needed, corporeality requires a violent death, and as soon as this is done, resurrection must take place in one form or another, for they go together. As Paul says, "If Christ be not risen, then is our preaching vain and your faith is also vain. . . . Ye are yet in sins."[1] The crucifixion in fact has a double sense: one individualistic and the other humanistic. In the first sense it symbolizes the destruction of the individual ego, while in the second it stands for the doctrine of vicarious

[1] I Cor., 15:14–17.

Crucifixion and Enlightenment

atonement whereby all our sins are atoned for by making Christ die for them. In both cases the dead must be resurrected. Without the latter, destruction has no meaning whatever. In Adam we die, in Christ we live—this must be understood in the double sense as above.

What is needed in Buddhism is enlightenment, neither crucifixion nor resurrection. A resurrection is dramatic and human enough, but there is still the odor of the body in it. In enlightenment, there are heavenliness and a genuine sense of transcendence. Things of earth go through renovation and a refreshing transformation. A new sun rises above the horizon and the whole universe is revealed.

It is through this experience of enlightenment that every being individually and collectively attains Buddhahood. It is not only a certain historically and definitely ascertainable being who is awakened to a state of enlightenment but the whole cosmos with every particle of dust which goes to the composition of it. I lift my finger and it illuminates the three thousand chiliocosms and an *asamkheyya* of Buddhas and Bodhisattvas greet me, not excluding ordinary human beings.

Crucifixion has no meaning whatsoever unless it is followed by resurrection. But the soil of the earth still clings to it though the resurrected one goes up to heaven. It is different with enlightenment, for it instantly transforms the earth itself into the Pure Land. You do not have to go up to heaven and wait for this transformation to take place here.

2

Christian symbolism has much to do with the suffering of man. The crucifixion is the climax of all suffering. Buddhists also speak much about suffering and its climax is the Buddha serenely sitting under the Bodhi tree by the river Niranjana. Christ carries his suffering to the end of his earthly life whereas Buddha puts an end to it while living and afterward goes on preaching the gospel of enlightenment until he quietly passes away under the twin Sala trees. The trees are standing upright and the Buddha, in Nirvana, lies horizontally like eternity itself.

Christ hangs helpless, full of sadness on the vertically erected cross. To the Oriental mind, the sight is almost unbearable. Buddhists are accustomed to the sight of Jizo Bosatsu (Kshitigarbha Bodhisattva) by the roadside. The figure is a symbol of tenderness. He stands upright but what a contrast to the Christian symbol of suffering!

Now let us make a geometric comparison between a statue sitting cross-legged in meditation and a crucified one. First of all, verticality suggests action, motion, and aspiration. Horizontality, as in the case of the lying Buddha, makes us think of peace and satisfaction or contentment. A sitting figure gives us the notion of solidity, firm conviction and immovability. The body sets itself down with the hips and folded legs securely on the ground. The center of gravity is around the loins. This is the securest position a biped can assume while living.

Crucifixion and Enlightenment 151

This is also the symbol of peace, tranquillity, and self-assurance. A standing position generally suggests a fighting spirit, either defensive or offensive. It also gives one the feeling of personal self-importance born of individuality and power.

When man began to stand on his two legs, this demonstrated that he was now distinct from the rest of the creatures walking on all fours. He is henceforth becoming more independent of the earth because of his freed forepaws and of the consequent growth of his brains. This growth and independence on the part of man are constantly misleading him to think that he now is master of Nature and can put it under his complete control. This, in combination with the Biblical tradition that man dominates all things on earth, has helped the human idea of universal domination to overgrow even beyond its legitimate limitation. The result is that we talk so much about conquering nature, except our own human nature which requires more disciplining and control and perhaps subjugation than anything else.

On the other hand the sitting cross-legged and the posture of meditation make a man feel not detached from the earth and yet not so irrevocably involved in it that he has to go on smelling it and wallowing in it. True, he is supported by the earth but he sits on it as if he were the crowning symbol of transcendence. He is neither attached to the soil nor detached from it.

We talk these days very much about detachment as if attachment is so fatal and hateful a thing that we must somehow try to achieve the opposite, non-attachment. But I do not know why we have to move away from things lovable and really conducive to our social and individual welfare. Kanzan and Jittoku enjoyed their freedom and welfare in their own way. Their life can be considered one of utter detachment as we the outsiders look at it. Śākyamuni spent his seventy-nine years by going from one place to another and teaching his gospel of enlightenment to all sorts of people varied in every way, social, intellectual, and economic, and finally passed away quietly by the river Niranjana. Socrates was born and died in Athens and used his energy and wisdom in exercising his office as the midwife of men's thoughts, bringing down philosophy from heaven to earth and finally calmly taking his cup of hemlock surrounded by his disciples and ending his life of seventy years.

What shall we say about these lives when each of them apparently enjoyed his to the utmost of his heart's content? Is it a life of attachment or of detachment? I would say that, as far as my understanding goes, each had his life of freedom unhampered by any ulterior interest and, therefore, instead of using such terms as attachment or detachment in order to evaluate the life of those mentioned above is it not better to call it a life of absolute freedom?

It is enlightenment that brings peace and freedom among us.

3

When Buddha attained his supreme enlightenment, he was in his sitting posture; he was neither

Crucifixion and Enlightenment

attached to nor detached from the earth. He was one with it, he grew out of it, and yet he was not crushed by it. As a newborn baby free from all *sankhāras*, he declared, standing, with one hand pointing to the sky and the other to the earth, "Above heaven, below heaven, I alone am the honored one!" Buddhism has three principal figures, symbolizing (1) nativity, (2) enlightenment, and (3) Nirvana, that is standing, sitting, and lying—the three main postures man can assume. From this we see that Buddhism is deeply concerned with human affairs in various forms of peaceful employment and not in any phase of warlike activities.

Christianity, on the other hand, presents a few things which are difficult to comprehend, namely, the symbol of crucifixion. The crucified Christ is a terrible sight and I cannot help associating it with the sadistic impulse of a physically affected brain.

Christians would say that crucifixion means crucifying the self or the flesh, since without subduing the self we cannot attain moral perfection.

This is where Buddhism differs from Christianity.

Buddhism declares that there is from the very beginning no self to crucify. To think that there is the self is the start of all errors and evils. Ignorance is at the root of all things that go wrong.

As there is no self, no crucifixion is needed, no sadism is to be practiced, no shocking sight is to be displayed by the roadside.

According to Buddhism, the world is the network of karmic interrelationships and there is no agent behind the net who holds it for his willful management. To have an insight into the truth of the actuality of things, the first requisite is to dispel the cloud of ignorance. To do this, one must discipline oneself in seeing clearly and penetratingly into the suchness of things.

Christianity tends to emphasize the corporeality of our existence. Hence its crucifixion, and hence also the symbolism of eating the flesh and drinking the blood. To non-Christians, the very thought of drinking the blood is distasteful.

Christians would say: This is the way to realize the idea of oneness with Christ. But non-Christians would answer: Could not the idea of oneness be realized in some other way, that is, more peacefully, more rationally, more humanly, more humanely, less militantly, and less violently?

When we look at the Nirvana picture, we have an entirely different impression. What a contrast between the crucifixion-image of Christ and the picture of Buddha lying on a bed surrounded by his disciples and other beings non-human as well as human! Is it not interesting and inspiring to see all kinds of animals coming together to mourn the death of Buddha?

That Christ died vertically on the cross whereas Buddha passed away horizontally—does this not symbolize the fundamental difference in more than one sense between Buddhism and Christianity?

Verticality means action, combativeness, exclusiveness, while horizontality means peace, tolerance, and broad-mindedness. Being active, Christianity has something in it which stirs, agitates, and disturbs. Being combative and exclusive, Christianity tends to wield an autocratic and sometimes dom-

Crucifixion and Enlightenment 155

ineering power over others, in spite of its claim to democracy and universal brotherhood.

In these respects, Buddhism proves to be just the opposite of Christianity. The horizontality of the Nirvana-Buddha may sometimes suggest indolence, indifference, and inactivity though Buddhism is really the religion of strenuousness and infinite patience. But there is no doubt that Buddhism is a religion of peace, serenity, equanimity, and equilibrium. It refuses to be combative and exclusive. On the contrary, it espouses broad-mindedness, universal tolerance, and aloofness from worldly discriminations.

To stand up means that one is ready for action, for fighting and overpowering. It also implies that someone is standing opposed to you, who may be ready to strike you down if you do not strike him down first. This is "the self" which Christianity wants to crucify. As this enemy always threatens you, you have to be combative. But when you clearly perceive that this deadly enemy who keeps you on the alert is non-existent, when you understand that it is no more than a nightmare, a mere delusion to posit a self as something trying to overpower you, you then will be for the first time at peace with yourself and also with the world at large, you then can afford to lie down and identify yourself with all things.

After all is said there is one thing we all must remember so as to bring antagonistic thoughts together and see how they can be reconciled. I suggest this: When horizontality remains horizontal all the time, the result is death. When verticality keeps up its rigidity, it collapses. In truth, the horizontal is horizontal only when it is coneived as implying the tendency to rise, as a phase of becoming something else, as a line to move to tridimensionality. So with verticality. As long as it stays unmoved vertically, it ceases to be itself. It must become flexible, acquire resiliency, it must balance itself with movability.

(The cross [Greek] and the swastika are closely related, probably derived from the same source. The swastika however is dynamic whereas the cross symbolizes static symmetry. The Latin cross is most likely the development of a sign of another nature.)

Section Two

chapter 7

Kono-mama ("I Am That I Am"[1])

I

The religious consciousness is awakened when we encounter a network of great contradictions running through our human life. When this consciousness comes to itself we feel as if our being were on the verge of a total collapse. We cannot regain the sense of security until we take hold of something overriding the contradictions.

Whatever contradictions we may experience they would not trouble us unless we were philosophers, because each one of us is not supposed to be a thinker of some kind. The contradictions however in most cases assert themselves in the field of the will. When we are assailed on this side, the question is felt most acutely, like a piercing arrow. When the will to power is exposed to constant threat in one form or another, one cannot help becoming meditative about life.

"What is the meaning of life?" then demands not an abstract solution but comes upon one as a concrete personal challenge. The solution must be in terms of experience. We then abandon all the contradictions that appear on the plane of intellection,

---

[1] "And God said unto Moses, I am that I am: and he said, Thus shalt thou say unto the children of Israel, I am hath sent one unto you." Exodus, 3:14.

for we must feel in a practical way contented with life.

The Japanese word *kono-mama* is the most fitting expression for this state of spiritual contentment. *Kono-mama* is the is-ness of a thing. God is in his way of is-ness, the flowers bloom in their way of is-ness, the birds fly in their way of is-ness—they are all perfect in their is-ness.

Christians ascribe all their ways of is-ness to God whoever he may be and remain satisfied with themselves in the midst of contradictions. John Donne (Sermon VII) has said: "God is so omnipresent . . . that God is an angel in an angel, and a stone in a stone, and a straw in a straw." Eckhart has his way of expressing the same idea: "A flea to the extent that it is in God, ranks above the highest angel in his own right. Thus, in God, all things are equal and are God himself." [2]

2

A Zen poet-master[3] sings:

> In the foreground precious stones and agates,
> In the rear agates and precious stones;
> To the East Kwannon and Seishi,
> To the West Monju and Fugen,[4]
> In the middle there is a streamer:

---

[2] Blakney, p. 205.
[3] Goso Hoyen, died 1104.
[4] These are the chief Mahāyāna Bodhisattvas:
  1. Avalokiteśvara
  2. Mahāsthāmaprāpta
  3. Mañjuśrī
  4. Samantabhadra

> As a breeze passes by
> It flutters, "*hu-lu,*" "*hu-lu.*"

This "*hu-lu,*" "*hu-lu*" (in Chinese) or "*fura-fura*" in Japanese reminds one of Saichi's outflowings:

1

> Saichi's mind is like the gourd [on water],
> Floating all the time,
> Blown by the winds, it flows on floating
> To Amida's Pure Land.

The one difference, we may point out, between Shin and Zen is that the Zen masters would not say "To Amedia's Pure Land." They would not mind if the gourd floats on to hell though they would not object to floating on to the Pure Land, either. This is not due to their indifference, *fura-fura*-ness. Superficially they may seem so, but only superficially. Their *fura-fura*-ness really comes from their deep experience of the Emptiness which concerns a life altogether transcendental or, we might say, "supernatural." Most people fail to distinguish the moral life from the inner transcendental life, which, it may be asserted, has a life of its own and lives altogether separate from an individually differentiated life which has its values in a world of utilitarian purposiveness.

To put all this again into Christian terminology, Eckhart declares:

> [If you can] take what comes to you through him, then whatever it is, it becomes divine in itself; shame becomes honor, bitterness becomes sweet, and gross darkness, clear light. Everything takes its flavor from God and be-

comes divine; everything that happens betrays God when a man's mind works that way; things all have this one taste; and therefore God is the same to this man alike in life's bitterest moments and sweetest pleasures.[5]

Eckhart naturally refers everything to God though his God somewhat resembles Saichi's "Namu-amida-butsu." We "creatures" as coming from God just follow his will *"sono-mama,"* and have nothing to say, good or bad, as to what we do. If I take this for the Christian understanding of *fura-fura*-ness, will Christians be offended?

Eckhart has a strange but interesting question in this connection:

A question is raised about those angels who live with us, serving and guarding us, as to whether or not they have less joy in identity than the angels in heaven and whether they are hindered at all in their [proper] activities by serving and guarding us. No! Not at all! Their joy is not diminished, nor their equality, because the angel's work is to do the will of God and the will of God is the angel's work. If God told an angel to go to a tree and pick off the caterpillars, the angel would be glad to do it and it would be bliss to him because it *is* God's will.[6]

The Shin pattern of expression is subjective and personal in contrast to the Zen way which is objective and impersonal, showing that Shin is more concerned with the *karuṇā* aspect of Reality while Zen tends to emphasize the *prajñā* aspect. The Shin

---
[5] Blakney, p. 17.
[6] *Ibid.*, p. 205.

Kono-mama ("I Am That I Am") 163

faith is based on Amida's *praṇidhāna*, which is summarized in the "Namu-amida-butsu" known as *myōgō* (*nāmadheya* in Sanskrit), meaning "the Name." The *myōgō* may sound abstract but it is the integrated form of subject and object, of devotee and Amida, of *Namu* (worshiper) and Buddha (the worshiped), of *ki* and *hō*.[7] When the *myōgō* is pronounced, the mystic identification takes place:

2

As I pronounce "Namu-amida-butsu"
I feel my thoughts and hindrances are like the spring snows:
They thaw away as soon as they fall on the ground.

3

Not knowing why, not knowing why—
This is my support;
Not knowing why—
This is the "Namu-amida-butsu."

4

Amida is this: "See, here I am!"
Namu and Amida—
They make out the "Namu-amida-butsu."
O Nyorai-san, such things I write,
How happy!

These are from Saichi. His experiences are given direct utterance here. As soon as the "Namu-amida-butsu" is pronounced he as *"Namu"* (*ki*) melts into the body of Amida (*hō*) which is "the ground" and

[7] See *infra*.

"support." He cannot reason it out, but "Here I am!" What has taken place is the identification of Amida (*hō*) and Saichi (*ki*). But the identification is not Saichi's vanishing. Saichi is still conscious of his individuality and addresses himself to Amida Buddha in a rather familiar fashion saying, "O Nyorai-san!" and congratulating himself on his being able to write about the happy event.

The following is the explosion of Mrs. Chiyono Sasaki of Kona on the island of Hawaii, who is a *myōkōnin* belonging to the Hongwanji Temple under Rev. Shōnen Tamekuni:

"*Kono-mamma*":[8] I am so pleased with this, I bow my head.
Good or bad—'tis "*kono-mamma*"!
Right or wrong—'tis "*kono-mamma*"!
True or false—'tis "*kono-mamma*"!
"Is" or "is not"—'tis '*kono-mamma*"!
Weep or laugh—'tis "*kono-mamma*"!
And "*kono-mamma*" is "*kono-mamma*"!

If you say "*kono-mamma*" is not enough, you are too greedy.
The "*kono-mamma*" never changes, nor can it be changed.
It is only because you are my Oya,
You call me to come "*sono-mama*" ["just as you are"].
It is all due to our not knowing that "*kono-mamma*" is "*kono-mama*"
That we wander about from one place to another.
That I am now inside the fold is due to the virtue of Oya's compassion,

[8] This is her colloquialism for *kono-mama*, "as-it-is-ness."

Kono-mama ("I Am That I Am")

And this pleases the Oya and also pleases me;
Oya and I live together then.
Each time I learn of his long-suffering labor,
How miserable I am!
How wretched I feel!
Ashamed of myself I resume my *Nembutsu:*
"Namu-amida-butsu, Namu-amida-butsu!"

"*Kono-mama,*" we may think, sounds too easy and there is nothing spiritual or transcendental in it. If we bring this out in the world of particulars, everything here will be left to the struggle for existence and the survival of the fittest. It is the most dangerous doctrine to be put forward, especially to our present world. But it may be worth asking, Is this doctrine of "*komo-mama*" really so dangerous?[9] Is our present world so valuable, deserving a careful preservation, where the knowledge of "that something" is fading away—"that something in the soul so closely akin to God," as Eckhart tells us? When

---

[9] Saichi has this to say in regard to "*kono-mama*":

> Though we all say by word of mouth "*kono-mama*"
> We really do not know what this means.
> There is no "*kono-mama*" to him who inattentively listens to the *Dharma,*
> There is no "*kono-mama*" to him.
> I am one of those heretics who know not what "*kono-mama*" means;
> A heretic indeed am I whose name is Saichi.

Saichi must have heard a preacher talking about "*kono-mama,*" warning the audience to be on their guard not to take it in the sense of indifference or dissipation or giving oneself to impulses of the moment which grow out of the one-sided self-power. The "*kono-mama*" doctrine is likely to turn into a "heresy" when it is only intellectually understood and not experienced in our innermost consciousness.

we did not have this knowledge, Eckhart regarded our individualistic ego-centered life "of no more importance than a manure worm." Is a world inhabited by this sort of existence really worth preservation?

Eckhart's passage is a very strong one:

> As I have often said, there is something in the soul so closely akin to God that it is already one with him and need never be united to him. It is unique and has nothing in common with anything else. It has no significance whatsoever—none! Anything created is nothing but that Something is apart from and strange to all creation. If one were wholly this, he would be both uncreated and unlike any creature. If any corporeal thing or anything fragile were included in that unity, it, too, would be like the essence of that unity. If I should find myself in this essence, even for a moment, I should regard my earthly selfhood as of no more importance than a manure worm.[10]

3

The doctrine of *"kono-mama"* is based on the psychology growing out of the experience of the Eckhartian "that something." One need not be metaphysically analytical in order to speak of it as eloquently as Eckhart does. But there is no doubt that Mrs. Sasaki, the author of the foregoing lines, tasted "that something" though she had no learning and mentality equal to the great German theologian. Saichi is more "learned" in his way and calls

---

[10] Blakney, p. 205.

Kono-mama ("I Am That I Am")

it "Buddha-wisdom" (*Buddhajñā*) which he must have heard from his preachers. Buddha-wisdom is really beyond our mere human understanding which is based on sensuous experiences and "logical" manipulations.

5

Buddha-wisdom is beyond thought,
Leading me to the Pure Land!
"Namu-amida-butsu!"

6

Perfectly indifferent I am!
No joy, no gratefulness,
Yet nothing to grieve over the absence of gratefulness.

7

Doing nothing, doing nothing, doing nothing!
Nyorai-san takes me along with him!
I am happy!

Saichi's "indifference" and "doing-nothing" are another and negative way of asserting *"kono-mama"* or *"sono-mama."* Buddha-wisdom is in one sense all-affirmation but in the other all-negation. It says "yes, yes" or *"sono-mama"* to everything that comes its way but at the same time it upholds nothing, saying: *"neti, neti."* When Saichi is in the negative mood, his bemoanings are "How wretched!" "How miserable!" "I am a sinner!" "I am a great liar." But when in the positive mood, everything changes. How jubilant he is! He is thankful for everything, he is most appreciative of Amida's free gift and wonders how he deserves it all. In spite of all this apparent fickleness or contradiction, Saichi keeps his mind well balanced and at peace, because his being is securely held in the hands of Amida and rests in the "Namu-amida-butsu."

8

Nothing is left to Saichi,
Except a joyful heart nothing is left to him;
Neither good nor bad has he, all is taken away from him;
Nothing is left to him!
To have nothing—how completely satisfying!
Everything has been carried away by the "Namu-amida-butsu."
He is thoroughly at home with himself:
This is indeed the "Namu-amida-butsu"!

9

"O Saichi."
"Yes, here I am."
"Where is your companion?"
"My companion is Amida-Buddha."
"Where are you?"
"I am in Amida."
"O Saichi."
"Yes."
"What is meant by 'all taken in and nothing left out'?"
"It means 'captured altogether'."
How grateful!
"Namu-amida-butsu!"

4

To clarify further Saichi's inner life, I quote more of his utterances:

Kōno-mama ("I Am That I Am")

10

All my cravings are taken away,
And the whole world is my "Namu-amida-butsu."

11

Saichi, all has been taken away from you,
And the *Nembutsu* is given—"Namu-amida-butsu!"

12

None of my evil passions, as many as 84,000, remain with me,
Every one of them has been taken away by the "Namu-amida-butsu."

13

My mind altogether taken captive by *bonnō*[11]
Has now been taken away together with *bonnō*.
The mind is enwrapped in the "Namu-amida-butsu"—
Thanks are due to the "Namu-amida-butsu"!

14

Saichi has nothing—which is joy.
Outside this there's nothing.
Both good and evil—all's taken away,
Nothing's left.
To have nothing—this is the release, this is the peace.
All's taken away by the "Namu-amida-butsu,"
This is truly the peace.
"Namu-amida-butsu!"

With all this, however, we must not think that Saichi turned into a piece of wood which is free

---

[11] Evil passions.

from all passions, good as well as bad. He was quite alive with them all. He was as human as we are ourselves. As long as we are what we are none of us can be released from the burden. To get rid of it means to get rid of our own existence which is the end of ourselves, the end of all things, the end of Amida himself, who has now no object for his *upāya* (or means) to exercise. The passions must remain with us all, without them there will be no joy, no happiness, no gratitude, no sociality, no human intercourse. Saichi is quite right when he asks Amida to leave *tsumi* with him;[12] he fully realizes that without *tsumi* he cannot experience Amida. Our existence is so conditioned on this earth that we must have one when we wish to have the other, and this wishing is no other than the passions which constitute *tsumi*. We are always involved in this contradiction, which is life, and we live it and by living it all is solved. The contradictions of any sort all turn into the *sono-mama*-ness of things. All that is needed is the experience of nothingness, which is suchness, *kono-mama*.

We can say that Saichi's free utterances, occupying more than sixty schoolchildren's notebooks, are rhapsodies on his living the grand contradiction itself which greets us as *ki* and *hō*[13] at every phase of

---

[12] See *infra*.

[13] *Ki*, originally meaning "hinge," means in Shin especially the devotee who approaches Amida in the attitude of dependence. He stands as far as his self-power is concerned against Amida. *Hō* is "*Dharma*," "Reality," "Amida," and "the other-power." This opposition appears to our intellect as contradiction and to our will as a situation implying anxiety, fear, and insecurity. When *ki* and *hō* are united in the

our existence. Saichi lives this contradiction and loves it to the fullest extent of his being. Each plane-shaving which rolls off his *geta*-making work table tells him about the world-drama which defies our attempt at solution on the plane of intellection. But the simple-minded Amida-intoxicated Saichi solves it quite readily by making each shaving bear his inscriptions: "What a despicable man I am!" and "How grateful Saichi is for Oya-sama's infinitely expanding compassionate heart!" "Poor Saichi is heavily *tsumi*-laden and yet he has no desire to part with it, for *tsumi* is the very condition that makes him feel the presence of Amida and his Namu-amida-butsu." Saichi lives the grand world-contradiction, his living is the solving.

5

Shinran, the founder of the Shin sect, presents this comment on the thought of "*kono-mama*." Since he was a Chinese scholar, he did not use the Japanese vernacular but its Chinese equivalent, *tzu-jên ja-êrh*, the Japanese reading of which is *jinen hōni*.

Ji means "of itself," or "by itself." As it is not due to the designing of man but to Nyorai's vow [that man is born in the Pure Land], it is said that man is naturally or

---

myōgō as "Namu-amida-butsu," the Shin devotee attains *anjin*, "peace of mind."

> O Saichi, if you wish to see Buddha,
> Look within your own heart where *ki* and *hō* are one
> As Namu-amida-butsu—
> This is Saichi's Oya-sama.
> How happy with the favor!
> Namu-amida-butsu, Namu-amida-butsu.

spontaneously (*nen*), led to the Pure Land. The devotee does not make any conscious self-designing efforts, for they are altogether ineffective to achieve the end. *Jinen* thus means that as one's rebirth into the Pure Land is wholly due to the working of Nyorai's vow-power, it is for the devotee just to believe in Nyorai and let his vow work itself out.

Hōni means "it is so because it is so"; and in the present case it means that it is in the nature of Amida's vow-power that we are born in the Pure Land. Therefore, the way in which the other-power works may be defined as "meaning of no-meaning," that is to say, it works in such a way as if not working [so natural, so spontaneous, so effortless, so absolutely free are its workings].

Amida's vow accomplishes everything and nothing is left for the devotee to design or plan for himself. Amida makes the devotee simply say "Namu-amida-butsu" in order to be saved by Amida, and the latter welcomes him to the Pure Land. As far as the devotee is concerned, he does not know what is good or bad for him, all is left to Amida. This is what I—Shinran—have learned.

Amida's vow is meant to make us all attain supreme Buddhahood. The Buddha is formless and because of his formlessness he is known as "all by himself" (*jinen*). If he had a form, he would not be called supreme Nyorai. In order to let us know how formless he is, he is called Amida. This is what I—Shinran—have learned.

When you have understood this, you need not any more be concerned with *jinen* ["being by itself"]. When you turn your attention to it, the "meaningless meaning" assumes a meaning [which is defeating its own purpose].

All this comes from *Buddhajñā*, which is beyond comprehensibility.

From this commentary of Shinran on *jinen hōni*,

Kono-mama ("I Am That I Am")

we can see what understanding he had of the working of Amida's *praṇidhāna* ("vows") or of the other-power. "Meaningless meaning" may be thought of as having no sense, no definite content, whereby we can concretely grasp what it means. The idea is that there was no teleology or eschatological conception on the part of Amida when he took those forty-eight "vows," that all the ideas expressed in them are the spontaneous outflow of his *mahākaruṇā*, great compassionate heart, which is Amida himself. Amida has no exterior motive other than a feeling of sorrow for us suffering sentient beings and a wish to save us from going through an endless cycle of births and deaths. The "vows" are the spontaneous expression of his love or compassion.

As for the sentient beings, they are helpless because they are limited existences, karma-bound, thoroughly conditioned by space, time, and causation. As long as they are in this state of finitude, they can never attain Nirvana or enlightenment by themselves. This inability to achieve emancipation is in the very nature of our existence. The more we try the deeper we get involved in an inextricable mess. The help has to come from a source other than this limited existence, but this source must not be something wholly outside us in the sense that it has no understanding of our limitations, and hence is not in any way sympathetic with us. The source of help must have the same heart as ours so that there will be a current of compassion running between the two. The source-power must be within us and yet outside. If not within us, it could not understand us; if not outside, it would be subject to the same conditions. This is an eternal problem—to be and not to be, to be within and yet to be outside, to be infinite and yet ready to serve the finite, to be full of meaning and yet not to have any meaning. Hence the incomprehensibility of *Buddhajñā*, hence the incomprehensibility of the "Namu-amida-butsu."

Saichi's version of Shinran has its own charm and originality from his inner experience which defies our logical analysis:

15

The Namu-amida-butsu inexhaustible,
However much one recites it, it is inexhaustible;
Saichi's heart is inexhaustible;
Oya's heart is inexhaustible.
Oya's heart and Saichi's heart,
Ki and *hō*, are of one body which is the Namu-amida-butsu.
However much this is recited, it is inexhaustible.

16

How wretched!—
This comes out spontaneously.
How grateful for Buddha's favor!—
This too spontaneously.
Ki and *hō*, both are Oya's working:
All comes out in perfection.

17

Saichi's Nyorai-san,
Where is he?
Saichi's Nyorai-san is no other than the oneness of *ki* and *hō*.
How grateful I am! "Namu-amida-butsu!"
"Namu-amida-butsu, Namu-amida-butsu!"

18

Such a Buddha! he is really a good Buddha!
He follows me wherever I go,
He takes hold of my heart.
The saving voice of the six syllables
Is heard as the oneness of *ki* and *hō*—
"Namu-amida-butsu!"
I have altogether no words for this;
How sweet the mercy!

Eckhart has his way of commenting on all these ideas which we may take as exclusively Shin.

> If you suffer for God's sake and for God alone, that suffering does not hurt and is not hard to bear, for God takes the burden of it. If an hundredweight were loaded on my neck and then someone else took it at once on his neck, I had just as lief it were an hundred as one. It would not then be heavy to me and would not hurt me. To make a long story short, what one suffers through God and for God alone is made sweet and easy.[14]

> It [the will] is perfect and right when it has no special reference, when it has cut loose from self, and when it is transformed and adapted to the will of God. Indeed, the more like this the will is, the more perfect and true it is. With a will like this, anything is possible, whether love or anything else.[15]

> Supposing, however, that all such [experiences] were really of love, even then it would not be best. We ought to get over amusing ourselves with such raptures for the sake of that better love, and to accomplish through loving service what men most need, spiritually or physically. As I have often said, if a person were in such a raptur-

---
[14] Blakney, p. 210.
[15] *Ibid.*, p. 13.

ous state as St. Paul once entered, and he knew of a sick man who wanted a cup of soup, it would be far better to withdraw from the rapture for love's sake and serve him who is in need.[16]

[16] *Ibid.*, p. 14.

Appendices

chapter 8

Notes on "Namu-amida-butsu"

The ultimate goal of the teaching of the Pure Land is to understand the meaning of *"Nembutsu,"* whereby its followers will be admitted into the Pure Land. In the *Nembutsu*, contradictions dissolve and are reconciled in "the steadfastness of faith."

Nembutsu literally means "to think of Buddha." *Nen* (*nein* in Chinese and *smṛiti* in Sanskrit) is "to keep in memory." In Shin however it is more than a mere remembering of Buddha, it is thinking his Name,[1] holding it in mind. The Name consists of six characters or syllables: *na-mu-a-mi-da-buts(u)* in Japanese pronunciation and *nan-wu-o-mi-to-fo* in Chinese. In actuality, the Name contains more than Buddha's name, for *Namu* is added to it. *Namu* is *namas* (or *namo*) in Sanskrit and means "adoration" or "salutation." The Name therefore is "Adoration for Amida Buddha," and this is made to stand for Amida's "Name."

The interpretation the Shin people give to the "Namu-amida-butsu" is more than literal though not at all mystical or esoteric. It is in fact philosophical. When Amida is regarded as the object of adoration, he is separated from the devotee standing all by himself. But when *Namu* is added to the Name the whole thing acquires a new meaning because it now symbolizes the unification of Amida

---

[1] *Myōgō, ming-hao* in Chinese, *nāmadheya* in Sanskrit.

and the devotee, wherein the duality no longer exists. This however does not indicate that the devotee is lost or absorbed in Amida so that his individuality is no longer tenable as such. The unity is there as "Namu" plus "Amidabutsu," but the *Namu* (*ki*) has not vanished. It is there as if it were not there. This ambivalence is the mystery of the *Nembutsu*. In Shin terms it is the oneness of the *ki* and the *hō*, and the mystery is called the incomprehensibility of Buddha-wisdom (*Buddhajñā*). The Shin teachings revolve around this axis of incomprehensibility (*fushigi* in Japanese, *acintya* in Sanskrit).

Now we see that the *Nembutsu*, or the *Myōgō*, or the "Namu-amida-butsu" is at the center of the Shin faith. When this is experienced, the devotee has the "steadfastness of faith," even before he is in actuality ushered into the Pure Land. For the Pure Land is no more an event after death, it is right in this *sahalokadhātu*, the world of particulars. According to Saichi, he goes to the Pure Land as if it were the next-door house and comes back at his pleasure to his own.

1

>   I am a happy man, indeed!
>   I visit the Pure Land as often as I like:
>   I'm there and I'm back,
>   I'm there and I'm back,
>   I'm there and I'm back,
>   "Namu-amida-butsu! Namu-amida-butsu!"

When Saichi is in the Pure Land, "there" stands for this world; and when he is in this world, "there" is the Pure Land; he is back and forth between here

Notes on "Namu-amida-butsu"

and there. The fact is that he sees no distinction between the two. Often he goes further than this:

2

How happy I am!
"Namu-amida-butsu!"
I am the Land of Bliss,
I am Oya-sama.
"Namu-amida-butsu! Namu-amida-butsu!"

3

Shining in glory is Buddha's Pure Land,
And this is my Pure Land!
"Namu-amida-butsu! Namu-amida-butsu!"

4

O Saichi, where is your Land of Bliss?
My Land of Bliss is right here.
Where is the line of division
Between this world and the Land of Bliss?
The eye[2] is the line of division.

To Saichi "Oya-sama" or "Oya" not only means Amida himself but frequently personifies the "Namu-amida-butsu." To him, sometimes, these three are the same thing: Amida as Oya-sama, the *Myōgō* ("Namu-amida-butsu"), and Saichi.

5

When I worship thee, O Buddha,
This is a Buddha worshiping another Buddha.
And it is thou who makest this fact known to me, O Buddha!
For this favor Saichi is most grateful.

---

[2] This reminds us of Eckhart.

When we go through these lines endlessly flowing out of Saichi's inner experiences of the "Namu-amida-butsu" as the symbol of the oneness of the *ki* and the *hō*, we feel something infinitely alluring in the life of this simple-minded *geta*-maker in the remote parts of the Far Eastern country. Eckhart is tremendous, Zen is almost unapproachable, but Saichi is so homely that one feels like visiting his workshop and watching those shavings drop off the block of wood.

6

O Saichi, what makes you work?
I work by the "Namu-amida-butsu."
"Namu-amida-butsu! Namu-amida-butsu!"

7

How grateful I feel!
Everything I do in this world—
My daily work for livelihood—
This is all transferred into building up the Pure Land.

8

I work in this world in company with all Buddhas,
I work in this world in company with all Bodhisattvas;
Protected by Oya-sama I am here;
I know many who have preceded me along this path.
I am sporting in the midst of the Namu-amida-butsu.
How happy I am with the favor!
"Namu-amida-butsu!"

To see Saichi work in the company of Buddhas and Bodhisattvas who fill up the whole universe[3] must

---

[3] Saichi has this:
   What a miracle!

be a most wonderfully inspiring sight. A scene transferred from the Pure Land! Compared with this, Eckhart appears to be still harboring something of this-worldliness. In Saichi all things come out of the mystery of the "Namu-amida-butsu" in which there is no distinction between "rapturous moments" and "love for one's neighbors."

There is another aspect in Saichi's life which makes him come close to that of a Zen-man. For he sometimes rises above the "Namu-amida-butsu," above the oneness of the *ki* and *hō*, above the ambivalence of wretchedness and gratefulness, of misery and joy. He is "indifferent," "nonchalant," "detached," or "disinterested" as if he came directly out of his "is-ness" in all nakedness, in the "*sono-mama*-ness" of things.

9

> Perfectly indifferent I am!
> No joy, no gratefulness!
> Yet no grief over the absence of gratefulness.

10

> "O Saichi, such as you are,
> Are you grateful to Amida?"
> "No particular feelings I have,
> However much I listen [to the sermons];
> And this for no reason."

At all events, Saichi was one of the deepest Shin followers, one who really experienced the mystery

---

> The "Namu-amida-butsu" fills the whole world!
> And this world is given me by Oya-sama!
> This is my joy!
> "Namu-amida-butsu!"

of the oneness of the *ki* and *hō* as symbolized in the "Namu-amida-butsu." He lived it every moment of his life, beyond all logical absurdities and semantic impossibilities.

11

O Saichi,[4] I am the most fortunate person!
I am altogether free from woes of all kind,
Not at all troubled with anything of the world.
Nor do I even recite the "Namu-amida-butsu"!
I'm saved by your mercifulness [O Amida-san!]
How pleased I feel for your favor!
"Namu-amida-butsu!"

12

While walking along the mountain path, how I enjoy smoking!
I sit by the roadside for a while, I take out the pipe in peace and with no trouble beclouding the mind.
But let us go home now, we have been out long enough, let us go home now.
How light my steps are as they move homeway!
My thoughts are filled with a return trip to Amida's country.
"Namu-amida-butsu, Namu-amida-butsu!"

---

[4] Saichi often begins his utterances with the self-addressing "O Saichi" and after a while forgets the way he started. "You" and "I" thus become confused. Here the grammatical niceties are disregarded.

chapter 9

Rennyo's Letters

"The letters" are those written by Rennyo Shōnin (1415-1499) to his followers. He was one of the greatest teachers of the Shin school of Buddhism; in fact it was he who laid the firm foundation for the modern religious institution known as the Jōdo-Shin Shu, the True Sect of the Pure Land. His letters numbering about eighty-five are preserved and the title *Gobunshō* or *Ofumi*, that is "honorable letters," is given to them. They are generally read before a sermon and quoted as the most authoritative source-documents on the teaching of the Shin school. Saichi states here that when his mind is illumined by these epistles he realizes what a miserable creature he is; but when he sees the Buddha-mind as revealed through these illuminating documents of the great teacher, he is assured of the overwhelming immensity of Oya-sama's love for him and feels grateful for it without measure. Rennyo's letters serve to bring out both aspects of our religious consciousness: (1) the sense of wickedness and depravity and (2) the feeling of gratitude for being saved from an utterly helpless situation. Here is one of Rennyo's letters:

To be established in [Shin] faith means to understand the Eighteenth Vow.[1] To understand the Eighteenth Vow

[1] "If upon my obtaining Buddhahood, all beings in the ten quarters should not desire in sincerity and trustfulness to be

means to understand the frame of mind[2] the "Namu-amida-butsu" sets up in you.

Therefore, when you attain a state of single-mindedness as you utter the "*Namu*" with absolute trust [in Amida] you perceive the significance of Amida's Vow which is directed towards awakening a faith-frame in you. For herein we realize what is meant by Amida Nyorai's "turning towards"[3] us ignorant beings. This is pointed out in the *Larger Sūtra of Eternal Life*[4] where we read: "Amida provides all beings with all the merits."[5]

Thus it follows that with all the evil deeds, with all the evil passions we have been cherishing in our former lives ever since the beginningless past, we are, owing to

---

in my country, and if they should not be born there by only thinking of me, say, up to ten times . . . may I not attain the Highest Enlightenment."

[2] As was explained, the *Nembutsu* which consists of the six syllables, *Na-mu-a-mi-da-buts(u)*, is a miraculous formula. When this is pronounced in sincerity of heart and in absolute faith in Amida, it produces a certain state of consciousness which is termed here "frame of mind" or "faith-frame." When it is attained, the devotee is said to have joined "the group of steadfastness" with no fear of retrogression. The faith thus awakened assures one of birth in the Pure Land.

[3] The miracle of Shin faith is that when the ordinary-minded people are confirmed in their faith all their sins and evil passions are transferred to Amida, and it is then he and not the actual sinners that would bear all the dire karma-consequences. More than that, all the merits Amida has accumulated during his infinite lives of self-training are given freely to the devotee. This is technically known as the doctrine of transference (*parināmanā*).

[4] This is the principal *Mahāyāna Sūtra* on which Shin teachings are based. But the Shin text is Sanghavarman's Chinese translation executed in 252 A.D., when he came to China from Central Asia.

[5] This "transference of merit" has a deep metaphysical significance in the history of Buddhist thought in India and China.

Rennyo's Letters 187

Amida's Vow which is beyond comprehension, thoroughly cleansed of them with no residue whatever left; and in consequence of it, we are made to abide with no fear of regression in "the order of steadfastness."[6]

This is what is meant by the statement that Nirvāna is attainable without destroying the evil passions (*kleśa*).[7]

This is the teaching exclusively taken up by our school but you are warned not to talk this way to people of other schools. Let me remind you of this.

With reverence . . .

The translation of such documents as Rennyo's letters is full of difficulties as they are so laden with technical terms which defy in many cases replacement by any other languages. The terms require lengthy explanations, which I have to omit. But a word about the "Namu-amida-butsu."

"Namu-amida-butsu" is the Japanese reading of the original Sanskrit phrase "namo amitābhabuddhāya," meaning "Adoration of the Buddha of Infinite Light." But with followers of the Pure Land teaching, the phrase is far more than mere adoration for Amitābhabuddha, or Amida, for by this they express their absolute faith in Amida as one who makes it possible for them to be born in his Land of Purity and Bliss.

With popular minds "Namu-amida-butsu" is

---

[6] The "order of steadfastness" is a stage where Shin devotees become absolutely sure of their rebirth in the Pure Land, that is to say, when they see as Saichi does all the doors removed which keep this world in separation from the Pure Land.

[7] *Nirvāna* is *kleśa* (*bonnō*) and *kleśa* is *Nirvāna*—this is one of the great Mahāyāna teachings. When however its import is not properly comprehended, it lends itself to all kinds of dangerous misinterpretation for which mysticism is usually blamed. Eckhart for this reason was censored as a heretic.

rather a confused notion, for as in the case of Saichi the phrase frequently represents Reality itself impersonated as Amida or Oya-sama, and at the same time it is a form of adoration as well as the expression of absolute dependence. This is not, however, all of "Namu-amida-butsu," for the phrase often serves as a metaphysical formula symbolizing the identity of subject and object, of the devotee and Amida, of the "sin-laden" individual and the all-saving and all-merciful Oya-sama, of all beings (*sarvasattva*) and Buddha, of *ki* and *hō*, of human yearnings and the supreme enlightenment. In this sense, the phrase, "Namu-amida-butsu," stands for a state of consciousness in which Saichi finds it sometimes difficult to distinguish himself from Amida.

13

> The Oya-sama who never fails me
> Has now become myself,
> Making me hear his Name—
> The "Namu-amida-butsu."

14

> I am a fortune one:
> Oya-sama is given me,
> The Oya who turns me into a Buddha—
> "Namu-amida-butsu!"

When the phrase is used as a philosophical symbol, it is usually divided first into two parts: "Namu" (*ki*) and "Amida-butsu" (*hō*). "Namu" then stands for the devotee filled with all possible sinfulness while "Amida-butsu" is the Buddha of infinite light and eternal life. When the devotee pronounces the phrase, "Namu-amida-butsu," he is the "Namu-amida-butsu" itself. When Saichi repeats "Namu-amida-butsu," "Namu-amida-butsu," the phrase is to be understood in this sense, and no idea of supplication or mere adoration is implied here. Saichi in this case may be said to be like Tennyson calling himself "Alfred," "Alfred" as he tells us in his "Ancient Sage." Saichi here is completely drunk with the identification, completely absorbed in true mystery, through which the miserable Saichi carrying all his human passions and cravings finds himself transformed into a Buddha and in the presence of all Buddhas and Bodhisattvas and other holy souls. In a state of ecstasy or intoxication, Saichi does not know where to stop when he jots down in his schoolboy's notebook all that goes through his mind while busying himself with making the footgear. Saichi's repetition of "Namu-amida-butsu" is to be interpreted in this way.

Saichi often addresses himself, asking such questions as: "What are you doing now, O Saichi?" "How are you faring, O Saichi?" "Say, Saichi, where are you?" "Why don't you stop writing?" and so on. These questions evidently show that his was a dual personality: the "miserable, despicable, woebegone" Saichi was living together with Amida or Oya-sama when Amida was felt to be near. Sometimes Saichi felt that it was not he who addressed Amida or himself but Amida addressing Amida. Amida's presence in Saichi was not a visionary experience. Amida really directed Saichi's movements while this by no means prevented Saichi from being himself, from being a miserable existence incalculably separated from Amida. But Saichi felt at the same moment that without this miserable existence of his he could not experience all the joy that came from unity.[8]

The psychologists may declare Saichi to be a very good example of schizophrenia. But they forget that Saichi is not a sick person, not a case of psychosis, tormented by the split. He is a perfectly healthy personality, he has never lost the sense of the oneness of his being. In fact his sense of being is so deep and yet so definite that he is living a more real and meaningful life than most of us do. It is we and not he who live in a psychological duality with all its disturbing consequences.

Eckhart once gave a sermon on "the just lives in eternity" in which he says:

The just lives in God and God in him, for God is born in the just and the just in God: at every virtue of the just God is born and is rejoiced, and not only every virtue but every action of the just wrought out of the virtue of the just and in justice; thereat God is glad aye, thrilled

---

[8] Meister Eckhart, as quoted by Paul Tillich in his paper on "The Types of Philosophy of Religion," says: "There is between God and the soul neither strangeness nor remoteness, therefore the soul is not only equal with God but it is—the same that He is." To Saichi, Amida is both remote and near, perhaps to him Amida is near because of his remoteness, he is remote because of his nearness. To use Saichi's own words: "Saichi feels miserable because he is full of evil cravings, but it is just because of his misery that he is made to appreciate the loving-kindness of Amida who is no other than his Oya-sama and that the joy and the feeling of gratitude following the appreciation know no bounds, even going beyond the limits of the whole universe."

with joy, there is nothing in his ground that does not dance for joy. To unenlightened (*grob*) people this is matter for belief but the illumined know.[9]

It is illuminating to hear Eckhart say that "to the coarse-minded (*grob*) people this is matter for belief but the enlightened know (*wissen*)." "The coarse-minded" means those who cannot go beyond the senses and the intellect, for they do not know anything that takes place in the realm of *prajñā*-intuition. The Oya-sama whose all-embracing and all-comprehending love makes Saichi hear his Name, "Namu-amida-butsu," first becomes Saichi himself. This means that the Oya-sama individualizes himself as a Saichi in the same way as Eckhart would have God "be born in the just and the just in God" and then hears his own Name pronounced by his individualized human Saichi. Amida is now transformed into the "Namu-amida-butsu" in the being of Saichi and Saichi in turn becomes Amida by hearing Amida's Name, "Namu-amida-butsu" as pronounced by Saichi himself. In this unity it is difficult to distinguish who is Amida and who is Saichi. When the one is mentioned the other inevitably comes along. Amida's Pure Land cannot now be anything else but Saichi's *sahaloka*—this *shaba* world of particular existences.

15

Oya is in the Pure Land,
I am in this world,
And Oya has given me,

[9] Evans, p. 149.

To become one with me:
The "Namu-amida-butsu"!

16

Let me go to the Pure Land,
Which is like visiting my neighbors—
This world is the Pure Land.
"Namu-amida-butsu!"

17

You are not saying the *Nembutsu*,
It is the *Nembutsu* that makes you say it,
And you are taken to the Pure Land.
"Namu-amida-butsu!"

chapter 10

From Saichi's Journals

The following are translations in English of some of Saichi's utterances. As I have said before, there are several thousands of such items in his journals, and there is no doubt that they are good material for students of religious experiences. My attempt here is, however poor the translations, to afford the reader a glimpse into Saichi's inner life. Unless one has a thorough mastery of both languages, Japanese and English, it is impossible to convey to the English reader the deep underlying feelings characterizing Saichi as one of the most conspicuously *myōkōnin* type of Shin followers.

The following selections, numbering 148, are grouped under nine headings. The classification is not at all scientific, since it is often very difficult to classify certain expressions under a certain definite group because they include various ideas interrelated to one another. The nine are as follows:

1. Nyorai and Saichi
2. Oya-sama
3. The *Nembutsu*
4. The *Ki* and the *Hō*
5. The Pure Land, This World and Hell
6. The Free Gift
7. The Heart-searchings
8. Poverty
9. The Inner Life

1. NYORAI[1] AND SAICHI[2]

1

I exchange work with Amida:
I worship him who in turn deigns to worship me—
This is the way I exchange work with him.[3]

2

"O Saichi, who is Nyorai-san?"
"He is no other than myself."
"Who is the founder [of the Shin teaching]?"
"He is no other than myself."
"What is the canonical text?"
"It is no other than myself."
The ordinary man's heart has no fixed root,
Yet this rootless one takes delight in the *Hō* [i.e., *Dharma*];

---

[1] *Nyorai* is the Japanese reading for Chinese *ju-lai*, which is the translation of the Sanskrit *tathāgata*. It means "one who thus comes (or goes)."

[2] *Cf.* Angelus Silesius, German mystic-poet:

> I know that without me
> God can no moment live;
> Were I to die, then He
> No longer could survive.
>
> I am as great as God,
> And He is small like me;
> He cannot be above,
> Nor I below Him be.

[3] The Japanese for "worship" is *ogamu*, which literally means "to bow to an object reverentially and devotionally." "Worship" may sound too strong, but if it is understood in the sense of "religious reverence and homage" as it is ordinarily done, there is no harm in the use of the term.

From Saichi's Journals

This is because he is given Oya's heart—
The heart of "Namu-amida-butsu."

3

I am lying.
Amida deigns to worship Saichi,
I too in turn worship Amida—
"Namu-amida-butsu!"

4

The adorable form of Nyorai
Is indeed this wretched self's form—
"Namu-amida-butsu, Namu-amida-butsu!" [4]

5

Buddha is worshiped by [another] Buddha:
The Namu is worshiped by Amida,
Amida is worshiped by the Namu:
This the meaning of *kimyo*[5]
As expressed in the "Namu-amida-butsu."

6

Amida calling on Amida—
This voice—
"Namu-amida-butsu, Namu-amida-butsu!"

---

[4] Dreamed on the night of May 22.

[5] *Kimyo* is the Japanese for *namu*, meaning "taking refuge," "adoration," "worshiping," etc. The author here probably intends to mean that mutual worshiping of Namu and Amida is the meaning of "Namu-amida-butsu," or that "Namu-amida-butsu" symbolizes the oneness of Amida and every one of us.

7

Saichi exchanges work with Amida:
When he worships Amida,
Amida in turn deigns to worship him [Saichi]—
This is the way we exchange our work.
How happy I am with the favor!
"Namu-amida-butsu, Nama-amida-butsu!"

8

When I worship thee, O Buddha,
This is Buddha worshiping [another] Buddha,
And it is thou who makest this fact known to me,
    O Buddha:
For this favor Saichi is most grateful.

9

What all the Buddhas of the *Hokkai*[6] declare
Is to make this Saichi turn into a Buddha—
"Namu-amida-butsu, Nama-amida-butsu!"

10

My joy!
How beyond thought!
Self and Amida and the "Namu-amida-butsu."

11

How fine!
The whole world and vastness of space is Buddha!
And I am in it—"Namu-amida-butsu!"

---

[6] *Hokkai* is *dharmadhātu* in Sanskrit, meaning the universe as the totality of all things.

2. OYA-SAMA [7]

12

Oya-sama is Buddha
Who transforms Saichi into a Buddha—
How happy with the favor!
"Namu-amida-butsu, Namu-amida-butsu!"

13

> My heart and Oya-sama—
> We have just one heart
> Of "Namu-amida-butsu."

14

I am a happy man,
A glad heart is given me;
Amida's gladness is my gladness—
"Namu-amida-butsu!"

15

The heart that thinks [of Buddha]
Is Buddha's heart,
A Buddha given by Buddha—
"Namu-amida-butsu!"

16

How grateful I am!
Into my heart has Oya-sama entered and fully
    occupies it.

---

[7] *Oya* has no English equivalent. It is both motherhood and fatherhood, not in their biological sense but as the symbol of loving-kindness. *Sama*, an honorific particle, is sometimes shortened to *san* which is less formal and more friendly and intimate.

The cloud of doubt all dispersed,
I am now made to turn westward.
How fortunate I am!
Saying "Namu-amida-butsu" I return west.

17

Are devils[8] come?
Are serpents come?
I know not.
I live my life embraced in the arms of Oya-sama,
I am fed with the milk of "Namu-amida-butsu,"
Looking at Oya-sama's face.
"Namu-amida-butsu!"

18

> When he is known as Oya,
> Worship him as such:
> Oya and I are one—
> The oneness of *ki* and *hō*
> In the "Namu-amida-butsu."

19

Amida is my Oya-sama,
I am child of Amida;
Let me rejoice in Oya-sama, in "Namu-amida-butsu."
The "Namu-amida-butsu" belongs to child as well as to Oya-sama:
By this is known the mutual relation [between Thee and me].

---

[8] *Oni* in Japanese, evil spirits under the King of Death (*yāmarāja*).

20

My heart and thy heart—
The oneness of hearts—
"Namu-amida-butsu!"

21

How lucky I am!
Oya is given me!
Oya who turns me into a Buddha is
The "Namu-amida-butsu!"

22

The *Hokkai* is my Oya—
Being my Oya—
"Namu-amida-butsu!"

23

Oya and child—
Between them not a shadow of doubt:[9]
This is my joy!

24

The *Namu* and Amida,
Oya and child,
They quarrel: the *Namu* on one side and Amida on
 the other side.
Repentance and joyfulness—
How intimate!

---

[9] Meaning absolute trust between Amida as Oya-sama and Saichi as child.

25

What is Saichi's understanding of the "Namu-amida-butsu"?
Yes, I am an adopted child of "Namu-amida-butsu."
How do you understand a life of gratitude?
As to being grateful, sometimes I remember it, sometimes I do not.
Really, a wretched man I am!

26

Namu-san[10] and Amida-san are talking:
This is the "Namu-amida-butsu" of Oya and son.

27

Namu-san and Amida-san—both are Amida:
"Namu-amida-butsu!"
This happiness is my happiness.

28

"Namu-amida-butsu!"—how grateful I am!
"Namu-amida-butsu!" is the oneness of the worldly and the highest truth.
"Namu-amida-butsu!"—how happy I am for the favor!
"Namu-amida-butsu, Namu-amida-butsu!"
Wherefrom is "Namu-amida-butsu"?
It is the mercy issuing from Oya's bosom;
How happy I am with the favor, "Namu-amida-butsu!"
"Wherefor is Saichi bound?"
"Saichi will go to the Land of Bliss."

---

[10] This is Saichi himself. *Namu* is personified here.

"With whom?"
"With Oya-sama I go—how happy I am!"
"Namu-amida-butsu, Namu-amida-butsu!"

3. THE NEMBUTSU[11]

29

"O Saichi, do you recite the *Nembutsu* only when you think of it?
What do you do when you do not think of it?"
"Yes, [well,] when I do not think of it, there is
The 'Namu-amida-butsu' [just the same]—
The oneness of *ki* and *hō*;
Even my thinking of [the *Nembutsu*] rises out of it.
How thankful I am for the favor!"
"Namu-amida-butsu, Namu-amida-butsu!"

30

Hōnen Shōnin [is said to have recited the *Nembutsu*] sixty thousand times [a day];
With Saichi it is only now and then.
Sixty-thousand-times and now-and-then—
They are one thing.
How grateful I am for the favor!
"Namu-amida-butsu!"

31

"O Nyorai-san, do you take me—this wretched one such as I am?

---

[11] The *Nembutsu* (literally "thinking of Buddha") and the *Myōgō* ("name") are often interchangeable. Both refer to the six syllables: *"Na-mu-a-mi-da-buts(u)."* The syllables serve three purposes: (1) as the *myōgō* itself, (2) as an actual invocation, and (3) as the symbol of identity.

Surely because of the presence of such wretched ones
 as you, Oya-sama's mercy is needed—
The Name is just meant for you, O Saichi,
And it is yours."
"That is so, I am really grateful,
I am grateful for the favor—
Namu-amida-butsu!"

32

All the miraculous merits accumulated by Amida
Throughout his disciplinary life of innumerable
 eons
Are filling up this body called Saichi.
Merits are no other than the six syllables "*Na-mu-a-mi-da-buts(u)*."

33

The "Namu-amida-butsu" is inexhaustible,
However much one recites it, it is inexhaustible;
Saichi's heart is inexhaustible;
Oya's heart is inexhaustible.
Oya's heart and Saichi's heart,
Ki and *hō*, are of one body which is the "Namu-amida-butsu."
However much this is recited, it is inexhaustible.

34

To Saichi such as he is, something wonderful has
 happened—
That heart of his has turned into Buddhahood!
What an extraordinary event this!
What things beyond imagination are in store within
 the "Namu-amida-butsu"!

35

The "Namu-amida-butsu"
Is like the sun-god,
Is like the world,
Is like the great earth,
Is like the ocean!
Whatever Saichi's heart may be,
He is enveloped in the emptiness of space,
And the emptiness of space is enveloped in "Namu-amida-butsu"!
O my friends, be pleased to hear the "Namu-amida-butsu"—
"Namu-amida-butsu" that will free you from *Figoku* [hell].

36

The *Nembutsu* is like vastness of space,
The vastness of space is illumined by Oya-sama's *Nembutsu*.
My heart is illumined by Oya-sama.
"Namu-amida-butsu!"

37

For what reason it is I do not know,
But the fact is the "Namu-amida-butsu" has come upon me.

38

How wretched! What shall I do?
[But] wretchedness is the "Namu-amida-butsu"—
"Namu-amida-butsu, Namu-amida-butsu!"

39

There is nothing in the *Hokkai;*
Only one there is,
Which is the "Namu-amida-butsu"—
And this is Saichi's property.

40

The "Namu-amida-butsu" is transformed and I am it,
And it delights in me,
And I am delighted in it.

41

How wretched!
And how joyous!
They are one
[In] the "Namu-amida-butsu."

42

The *Nembutsu* of repentance over my wretchedness,
The *Nembutsu* of joy—
The "Namu-amida-butsu."

43

I may be in possession of 84,000 evil passions,
And Amida too is 84,000—
This is the meaning of oneness of "Namu-amida-butsu."

44

The *Namu* is myself,
Amida is the *Namu;*
And both *Namu* and Amida are the "Namu-amida-butsu."

45

I, bound for death,
Am now made into the immortal "Namu-amida-butsu."

46

Life's ending means not-dying;
Not-dying is life's ending;
Life's ending is to become "Namu-amida-butsu."

47

Death has been snatched away from me,
And in its place the "Namu-amida-butsu."

48

Saichi's heart destined for death when his end comes,
Is now made an immortal heart,
Is made into the "Namu-amida-butsu."

49

To die—nothing is better than death;
One feels so relieved!
Nothing exceeds this feeling of relief.
"Namu-amida-butsu, Namu-amida-butsu!"

4. THE KI AND THE HŌ[12]

50

"O Saichi, let me have what your understanding is."
"Yes, yes, I will:

[12] The following equations hold: the *Ki* = *Jiriki* ("self-power") = the *Namu* = the supplicating individual = the sin-

How miserable, how miserable!
Namu-amida-butsu, Namu-amida-butsu!"
"Is that all, O Saichi?
It will never do."
"Yes, yes, it will do, it will do.
According to Saichi's understanding,
Ki and *hō* are one:
The 'Namu-amida-butsu' is no other than he himself.
This is indeed Saichi's understanding:
He has flowers in both hands,
Taken away in one way and given as gift in another way."

51

How happy I am for this favor! "Namu-amida-butsu!"
Now I know where to deposit all my amassed delusions:
It is where the *ki* and the *hō* are one—
The "Namu-amida-butsu."

52

Such a Buddha! he is really a good Buddha!
He follows me wherever I go,
He takes hold of my heart.
The saving voice of the six syllables
Is heard as the oneness of the *ki* and the *hō*—
As the "Namu-amida-butsu."

---

ner = Saichi. The *Hō* = Amida = Buddha = Enlightenment = *Tariki* ("other-power") = Reality = the *Dharma* = Oya-sama = Tathāgata.

I have altogether no words for this;
How sweet the mercy!

53

No clinging to anything (*kata-giru ja nai*):
No clinging to the *ki*,
No clinging to the *hō*—
This is in accord with the Law (*okite ni kanō*).
"Namu-amida-butsu!"

This on the part of the *ki*,
This on the part of the *hō*.
How grateful I am!
"Namu-amida-butsu!"

54

How wretched!
What is it that makes up my heart?
It is no other than my own filled with infinitude of guilt,
Into which the two syllables *na-mu* have come,
And by these syllables infinitude of guilt is borne,
It is Amida who bears infinitude of guilt.
The oneness of the *ki* and the *hō*—
"Namu-amida-butsu!"

55

Saichi's Nyorai-san,
Where is he?
Saichi's Nyorai-san is no other than the oneness of the *ki* and the *hō*.
How grateful I am! "Namu-amida-butsu!"
"Namu-amida-butsu, Namu-amida-butsu!"

56

O Saichi, if you wish to see Buddha,
Look within your own heart where the *ki* and the *hō* are one
As the "Namu-amida-butsu"—
This is Saichi's Oya-sama.
How happy with the favor!
"Namu-amida-butsu, Namu-amida-butsu!"

57

If the *Namu* is myself, Amida is myself too:
This is the "Namu-amida-butsu" of six syllables.[13]

58

The *Namu* is worshiped by Amida,
And Amida is worshiped by the *Namu*—
This is the "Namu-amida-butsu" of six syllables.

59

This Sachi is thine,
Thou art mine—
"Namu-amida-butsu!"

60

As to Saichi's own Nyorai-san
Where is he?
Yes, Saichi's Nyorai-san is the oneness of the *ki* and *hō*.

---

[13] Saichi generally declares *Namu* to be himself and Amida to be Oya-sama. To identify himself with both *Namu* and Amida is unusual. We may however remark that Saichi often equates himself with "Namu-amida-butsu," which means that he is Amida as well as *Namu*.

From Saichi's Journals

How grateful I am!
"Namu-amida-butsu!"
"Namu-amida-butsu!"

61

"O Saichi, what are you saying to Oya-sama?"
"I am saying, 'Amida-bu, Amida-bu'."
"What is Oya-sama saying?"
"He is saying, '*O Namu, O Namu*'."
Thus Thou to me, and I to Thee:
This is the oneness of the *ki* and the *hō*.
"Namu-amida-butsu!"

62

"O Saichi, how do you see 'thee'?"
"To see 'thee' [take] Amida's mirror,
Therein revealed are both *ki* and *hō*.
Beyond that—repentance and joy.
How wonderful, how wonderful!
Grateful indeed I am! Namu-amida-butsu!"

63

How wretched!—
This comes out spontaneously.
How grateful for Buddha's favor!—
This too spontaneously.
The *ki* and the *hō*, both are Oya's working.

64

All comes out in perfection.
How grateful for the favor!
And I take no part in it.
How grateful for the favor!
"Namu-amida-butsu, Namu-amida-butsu!"

5. THE PURE LAND, THIS WORLD AND HELL[14]

65

"O Saichi, what is your pleasure?"
"My pleasure is this world of delusion;
Because it turns into the seed of delight in the
    Dharma (*hō*)."
"Namu-amida-butsu, Namu-amida-butsu!"

66

This world (*sahaloka*) and the Pure Land—they are
    one;
Worlds as numberless as atoms, too, are mine.
"Namu-amida-butsu, Namu-amida-butsu!"

67

The path to be born into the Land of Bliss
From this world, there is no other, after all,
Than this world itself.
This world is Namu-amida-butsu
Just as much as the Land of Bliss is.
How grateful, how grateful I am!
This Saichi's eye[15] is the boundary line

---

[14] *Jigoku* is hell generally, *Gokuraku* is the Land of Bliss, *Jōdo* is the Pure Land, and *shaba* is "this world" or *sahalokadhātu* in Sanskrit.

[15] This does not necessarily mean that when the eyes are closed which symbolize death we are in the Pure Land and that while they are kept open we are in this world. Saichi's idea probably is metaphysical or dialectical, though of course this is not to say that Saichi has reasoned out all these things consciously after the fashion of a philosopher. Saichi's allusion to the eye reminds us of Eckhart's remark on it.

[Between this world and the Land of Bliss].
"Namu-amida-butsu, Namu-amida-butsu!"

68

Where are you sleeping, O Saichi?
I am sleeping in this world's Pure Land;
When awakened I go to Amida's Pure Land.

69

This is *shaba* (Sanskrit: *sahaloka*),
And my heart is born of *Figoku* (Sanskrit: *naraka*).

70

"O Saichi, when you die, who will be your companion to the Land of Bliss?"
"As to me, Emma-san will be my companion."
"O Saichi, you tell us such tales again.
Who has ever gone to the Land of Bliss with Emma-san as companion?
O Saichi, you'd better not tell us such nonsense any more."
"In spite of your remark, I say you are mistaken;
Have you not read this in the 'Songs'?
'Emma, Great Lord of Justice, respects us; together with lords of the five paths, he stands as guardian day and night.'
You too should rejoice in the company of Emma-sama—
Here is Namu-amida-butsu.
This world, how enjoyable with Emma-sama!
This Saichi too is guarded by Emma-sama,
This Saichi and Emma-sama both are one Namu-amida-butsu:

This is my joy!"
"O Saichi, from whom did you get such a joyous note?"
"Yes, I talked with Emma-sama himself who granted this to me—
[He says] 'You are welcome indeed.'
How joyful! how joyful!
Namu-amida-butsu! Namu-amida-butsu!"

71

> I'm fortunate indeed!
> Not dead I go,
> Just as I live,
> I go to the Pure Land!
> "Namu-amida-butsu!"

72

> Led by "Namu-amida-butsu,"
> While living in this world,
> I go to "Namu-amida-butsu."

73

> This *shaba* world
> Is a little court-yard in the Pure Land;
> Looking at the court-yard
> I enjoy the Pure Land—
> "Namu-amida-butsu!"

74

> I am poor and immensely happy at that;
> Amida's Pure Land I enjoy while here—
> "Namu-amida-butsu!"

75

If the *shaba* world is different from the Pure Land,
I should never have heard the *Dharma*:

From Saichi's Journals

Myself and this *shaba* world and the Pure Land and
 Amida—
All is one "Namu-amida-butsu."

76

This *shaba* world too is yours,
Where Saichi's rebirth is confirmed—
This is your waiting teahouse.

77

This *shaba* turned into the Pure Land,
And myself changing!
"Namu-amida-butsu!"

78

My joy is that while in this world of *shaba*
I have been given the Pure Land—
"Namu-amida-butsu!"

79

My birthplace? I am born of *Figoku* (hell);
I am a nobody's dog
Carrying the tail between the legs;
I pass this world of woes,
Saying "Namu-amida-butsu."

80

How happy I am! "Namu-amida-butsu!"
I am the Land of Bliss;
I am Oya-sama.
"Namu-amida-butsu!"

81

Shining in glory is Amida's Pure Land,
And this is my Pure Land—
"Namu-amida-butsu!"

82

Heard so much of the Happy Land,
But after all it is not so much [as I expected];
It is good that it is not, indeed,
How at home do I feel with it!
"Namu-amida-butsu, Namu-amida-butsu!"

83

The Land of Bliss is mine,
Just take "Namu-amida-butsu" as you hear it!

84

How grateful!
While others die,
I do not die:
Not dying, I go
To Amida's Pure Land.

85

Has Saichi ever seen the Land of Bliss?
No, Saichi has never seen it before.
This is good—
The first visit this.

86

How grateful I am!
I live without knowing anything—
Is this living in a natural Pure Land?

87

How grateful I am!
Into my heart has Oya-sama entered!

From Saichi's Journals

The cloud of doubt is all dispersed,
I am now given to turn westward.
How fortunate I am!
Saying "Namu-amida-butsu," I turn west.

88

Buddha-wisdom is beyond human thought,
It makes me go to the Pure Land.
"Namu-amida-butsu!"

89

How dreadful!
This world known as *shaba*
Is where we endlessly commit all kinds of karma.
How thankful!
All this is turned into [the work of] the Pure Land,
Unintermittently!

90

The most wonderful thing is
That Buddha's invisible heart of compassion is visible
While I'm right here;
That the Pure Land, millions of millions of worlds away, is visible
While I'm right here—
"Namu-amida-butsu!"

91

I am not to go to *Figoku* (hell),
Figoku is right here,
We are living right in *Figoku*,
Figoku is no other place than this.

92

The *Hokkai* is never filled
However much we may talk of it—
Which is the Land of Bliss.
"Namu-amida-butsu!"

93

The *Hokkai* is Saichi's own country—
"Namu-amida-butsu!"

94

There is a man going back to Amida's Pure Land—
The *Namu* is carried by Amida.
The Pure Land where he returns
Is the "Namu-amida-butsu."

95

The being reborn means this present moment;
By means of the "Namu-amida-butsu" this is attained;
"Namu-amida-butsu!"

6. THE FREE GIFT[16]

96

Let this world go as it does,
Ignorance-debts, all paid up by Nyorai-san—
How happy, how happy I am!

---

[16] The gift or favor coming from Amida is a free one, for he never asks anything in exchange or in compensation. When the sinner (*ki*) utters "Namu-amida-butsu" in all sincerity he is at once made conscious of his being from the first with

97

Whatever we might say, it is all from thy side,
Yes, it is all from thee.
How thankful I am indeed, how happy I am indeed!
"Namu-amida-butsu, Namu-amida-butsu!"

98

The "Namu-amida-butsu" is as great as the world itself;
All the air is the "Namu-amida-butsu";
My heart is also a big heart,
My *tsumi* is filling the world.
However bad Saichi may be, he cannot defeat you, [O Buddha];
My *tsumi* is dragged along by you,
And it is now taken up [by you] to the Pure Land—
This favor of yours, this favor of yours!
"Namu-amida-butsu!"

99

The treasure of the six syllables was given me by Oya-sama:
However much one spends of it, it is never exhausted.
The treasure grows all the more as it is used;
It is the most wondrous treasure,
And I am the recipient of the good thing.

Amida and in Amida. There has never been any sort of alienation or estrangement between Amida and sinner. It was all due to the latter's illusive ideas cherished about himself. When they are wiped away, he realizes that the sun has always been there and finds himself basking in its light of infinity.

How happy I am with the favor! "Namu-amida-butsu!"

100

"O Saichi, you say 'I am given, I am given'
And what is it that is given you?"
"Yes, yes, I am given, I am given the Name of Amida!
And this for nothing!
Saichi is thereby set at ease.
To be set at ease means that the *ki* is altogether possessed [by Oya-sama].
It is indeed Oya-sama who has taken full possession of me,
And this Oya-sama of mine is the 'Namu-amida-butsu.'"

101

Saichi has his heart revealed by Amida's mirror,
How happy for the favor! "Namu-amida-butsu!"
"Namu-amida-butsu, Namu-amida-butsu!"
"Namu-amida-butsu, Namu-amida-butsu!"

102

What a miracle! The "Namu-amida-butsu" fills up the whole world,
And this world is given to me by Oya-sama.
This is my happiness, "Namu-amida-butsu!"

103

O Nyorai-san,
You have given up yourself to me,

From Saichi's Journals

And my heart has been made captive by you—
"Namu-amida-butsu!"

104

How miserable!
Saichi's heart, how miserable!
All kinds of delusion thickly arise all at once!
A hateful fire mixed with evils is burning,
The waves mixed with evils are rising,
How miserable! A fire mixed with follies is burning.

This heretic, how miserable!
Cannot you call a halt?

Saichi's heart, worrying,
A heart in utter confusion,
Saichi's heart rising as high as the sky!
Here comes the wise man giving the warning:
"O Saichi, listen, now is the time!"

How grateful!
"Now that Amida's 'Original Vow' is established as the 'Namu-amida-butsu,'
You have no more to worry about yourself,
Listen, listen!
When you hear 'Namu-amida-butsu,'
You have your rebirth in the Pure Land.
The 'Namu-amida-butsu' is yours."

How happy I am for this favor! "Namu-amida-butsu!"
Now I know where to deposit all my amassed delusions:

It is where the *ki* and *hō* are one—
The "Namu-amida-butsu."
With this heart [thus identified],
All over the worlds as many as atoms,
I roam playing in company with all Buddhas and Bodhisattvas.

Eating the "Namu-amida-butsu," this heart passes its time
In happy company with the "Namu-amida-butsu."

How happy with this favor!
"Namu-amida-butsu!"

105

O you, my friends, looking at your hearts filled with wretchedness,
Be not led to doubt Amida's mercy,
Though there is indeed this possibility.
But this is the greatest mistake you are apt to commit.
An utter wretchedness we all guilty beings experience
Does surely turn into a priceless treasure—
This you will realize when karma ripens;
For the "Namu-amida-butsu" truly achieves wonders.

That the "Namu-amida-butsu" truly achieves wonders is this:
The oceans, mountains, eatables, waters, wood used for our house-building, and all other things handled by us guilty beings:

From Saichi's Journals

They are one and all transformations of the "Namu-amida-butsu."
O my friends, be pleased to take note of this truth,
For this is all due to Oya-sama's mercy.
How grateful I feel for all this!

"Namu-amida-butsu, Namu-amida-butsu!"

106

How grateful!
When I think of it, all is by his [Amida's] grace.
O Saichi, what do you mean by it?
Ah, yes, his grace is real fact.
This Saichi was made by his grace;
The dress I wear was made by his grace;
The food I eat was made by his grace;
The footgear I put on was made by his grace;
Every other thing we have in this world was all made by his grace,
Including the bowl and the chopsticks;
Even this workshop where I work was made by his grace:
There is really nothing that is not the "Namu-amida-butsu."
How happy I am for all this!
"Namu-amida-butsu!"

107

By your favor I am turned into a Buddha;
Infinitely great is this favor of yours—
"Namu-amida-butsu!"

108

"Saichi's illness, is it cured by swallowing the 'Namu-amida-butsu'?"
"O, no!"
"If so, how is it cured?"
"Yes, Saichi's illness is cured when it is swallowed up by the 'Namu-amida-bu-sama.'"
Saichi is now bodily swallowed up by the pill of the six syllables,
And within the six syllables he leads a life of gratitude.
His life of gratitude is indeed a mystery,
The mystery of mysteries this!
How happy I am with the favor!
"Namu-amida-butsu!"

109

Saichi has something good given him,
The meditation of five Kalpas is given him.
Where can he have a fit place to store such a big thing?
The fact is that he is taken into it.
How grateful I am!
"Namu-amida-butsu!"
"Namu-amida-butsu!"

110

Saichi has something good given him,
The meditation of five Kalpas is given him.
Where can he have a fit place to store such a big thing?

The fact is that he is taken into it.
How grateful I am!
"Namu-amida-butsu!"
"Namu-amida-butsu!"

111

"Namu-amida-butsu" is indeed a wonderful Name,
And I have it as gift.
It gushes out of Saichi's heart;
This is as it ought to be:
The *ki* and *hō* are one in the "Namu-amida-butsu."

112

"O Saichi, tell us what kind of taste[17] is the taste of 'Namu-amida-butsu,'
Tell us what kind of taste is the taste of 'Namu-amida-butsu.' "
"The taste of the 'Namu-amida-butsu' is:
A joy filling up the bosom,
A joy filling up the liver,
Like the rolling swell of the sea—
No words—just the utterance: Oh, Oh!"

113

There is one thing I wish to learn from Oya-sama:
How do you wipe out my guilts?
Carrying my guilts as they are

---

[17] "Taste"—Bible reference: *Imitation of Christ*, Chapter XXXIV. "To him who tasteth Thee, what can be distasteful? And to him who tasteth Thee not, what is there which can make him joyous?"

[I am] borne up by the "Namu-amida-butsu"!
How grateful I am!
"Namu-amida-butsu, Namu-amida-butsu!"

114

The three poisonous passions are in company with the "Namu-amida-butsu,"
And found working with "Namu-amida-butsu"!
How thankful I am for the favor!
"Namu-amida-butsu!"

115

The love that inspired Oya-sama to go through
All the sufferings and all the hardships—
I thought I was simply to listen to the story,
But that was a grievous mistake, I find.

116

[What a wonder] that such a bad man as Saichi whose badness knows no bounds
Has been transformed into a Buddha!
How grateful for the favor, and how happy!
"Namu-amida-butsu, Namu-amida-butsu!"

117

How wretched I am!
For us ordinary people human calculations are of no avail.
As to the estimation of guilts—this is left to Oya-sama.
How grateful for the favor!
"Namu-amida-butsu!"

118

My heart given up to Thee.
And Thy heart received by me!

7. THE HEART-SEARCHING[18]

119

The *bombu*[19] cannot live with Buddha,
Because he has no humility and joy;
Lives with Buddha—
"Namu-amida-butsu!"

120

Saying "I cannot understand,"
They seize upon the *bonnō*[20] and investigate;

---

[18] While at the moment of exaltation Saichi feels he is Amida himself in company with Buddhas and Bodhisattvas who fill the whole universe, there are occasions when he feels the contrary. He then is the most despicable creature, like a homeless dog with his tail between the legs. He would cry: "How wretched, how worthless, how full of 84,000 evil thoughts am I!" But he never remains long in this state of self-commiseration, for he soon rises from it triumphantly, praising Buddha's infinite love for him. The psychologist may take him as a good example of manic-depressive psychosis. But the trouble is that Saichi is very much saner than most ordinary minds including scholars. He belongs to the group of "steadfastness," he has "something" occupying the very core of his being as Eckhart would say. Students of the religious consciousness know well that there is something of ambivalence in every devout soul. In this respect Saichi's utterances are of unusual importance.

[19] *Bombu* is the unenlightened and stands in contrast to Buddha.

[20] *Kleśa* in Sanskrit, generally rendered "evil passions." They are the product of ignorance (*avidyā*) and thirst (*tṛiṣṇā*).

But the *bonnō* is the body of merit;
This makes me laugh.

121

If there were no wretchedness,
My life would be wickedness itself;
How fortunate I am that I was given wretchedness
"Namu-amida-butsu, Namu-amida-butsu!"

122

When the *bombu* is not understood
It is wickedness;
When understood, it is humility—
"Namu-amida-butsu!"

123

Saichi feels within himself
An endless flow of folly,
An endless flow of greed;
There is a fire constantly burning—
No wonder, this burning,
For Saichi is an evil spirit.

124

Saichi's heart is all rain,
Saichi's heart, like rain and rain, is all rain;
Saichi's heart is all fog, like fog within a fog.
There is nothing but wretchedness in Saichi's heart.

125

"How wretched I am!"
This is what we all say when we feel humiliated.

From Saichi's Journals

But this kind of self-humiliation we say now is all a lie.
[The real one we say is] what we say after we've visited the Pure Land.
This Saichi's self-humiliation is nothing but a lie, monstrous lie, a monstrous, monstrous lie!
And within this lie there is another lie well wrapped!
How shameful!
This "How shameful!" is also a lie bursting out of the mouth.
This Saichi putting on the mask is most irreverently playing upon the saintly masters!
How wretched, how wretched!
There, there, that Saichi is again putting on the mask!
There is nothing in this Saichi but going around in disguise and deceiving everybody;
How wretched!
Anything Saichi says is wretchedness itself.
Even this comes out of the lying lips.
The only real true thing is Oya-sama, no other there is!
All my lies have been completely taken away [by him],
[And there remains nothing but]
"Namu-amida-butsu!"

126

How did you see your own heart?
To see the heart, take Amida's mirror.
How wretched!

The wretchedness of my heart is like space, it has no
    limits.
How wretched!

127

O Saichi, you are a wretched fellow!
Your stature is hardly five feet,
And yet your heart runs wildly all over the world.
Saichi is a wretched man.
How wretched!

128

They understand who have had sorrows,
But those who had them not can never understand:
There is nothing so excruciating as sighs—
The sighs that refuse to be disposed of.
But they are removed by Amida,
And all I can say now is "Namu-amida-butsu,
    Namu-amida-butsu!"

129

There is no bottom to Saichi's wickedness
There is no bottom to Saichi's goodness:
How happy I am with the favor!
"Namu-amida-butsu!"

130

The wretched heart of contrition—
The thankful heart of joy—
The "Namu-amida-butsu" of contrition and joy! [21]

---

[21] Logically speaking, this is a case of identity in absolute contradiction. Saichi demonstrates this experientially. When

8. POVERTY[22]

131

Nothing is left to Saichi,
Except a joyful heart nothing is left to him.
Neither good nor bad has he, all is taken away from
 him;
Nothing is left to him!
To have nothing—how completely satisfying!
Everything has been carried away by the "Namu-
 amida-butsu."
He is thoroughly at home with himself:
This is indeed the "Namu-amida-butsu"!

132

My avarice has all been taken away,
And the world has turned into my "Namu-amida-
 butsu."

133

Everything of mine has been carried away by Thee,
And Thou hast given me the *Nembutsu*—"Namu-
 amida-butsu."

---

he is conscious of his finiteness, being bound to the law of karmic causation, his heart is filled with contrition. But as soon as he feels that it is because of this consciousness that he has been taken up in the arms of Oya-sama, his joy knows no limits. The "Namu-amida-butsu" symbolizes the unification or rather identification of utter wretchedness and elated joyfulness.

[22] Poverty means that all that one thinks to be one's own is taken or carried away by Amida or Oya-sama, that the self-power (*Jiriki*) finds itself of no avail whatever. More positively, it is a state of self-realization that Amida is all in all.

9. THE INNER LIFE[23]

134

To be grateful is all a lie,
The truth is—there is nothing the matter;
And beyond this there is no peace of mind—
"Namu-amida-butsu, Namu-amida-butsu, Namu-amida-butsu!"
(With this peacefully I retire.)

135

There's nothing with me, nothing's the matter with me—
To have nothing the matter is the "Namu-amida-butsu."

136

That this Saichi is turned into a Buddha,
Even while I knew nothing of it:
So I am told.

137

How wretched!
Wretchedness too is of suchness.[24]

---

[23] The inner life is the life of suchness, of *kono-mama*, of the "nothing's the matter," of the "I know not what," of the horse galloping on the heath (Eckhart), of the flea in God's is-ness.

[24] The original Japanese reads, "*onodzukara*," which means "as-it-is-ness," "being natural," "being perfect in itself," or "being sufficient in itself." This is *kono-mama* or *sono-mama*.

How thankful!
Buddha's favor too is of suchness.
Both *ki* and *hō* are Oya-sama's work.[25]
All out, nothing kept back! [26]
How grateful for the favor!
Nothing's left for me to do.
How grateful for the favor!
"Namu-amida-butsu!"

138

As regards myself, nothing is the matter:
Called by the voice the mind has been made captive,
And "Namu-amida-butsu!"

139

To say, "How grateful!" is a lie;
The truth is: there is nothing the matter with one;
And there is nothing more that makes one feel at home—
"Namu-amida-butsu! Namu-amida-butsu!"

140

O Saichi, such as you are, are you grateful?
Nothing's the matter [with me],
However much I listen [to the sermons], nothing's the matter with me.
And no inquiries are to be made.

---

[25] *Hataraki* in the original means "function," "action," or "operation."

[26] *Marude deru*, meaning "to come out in all nakedness," "nothing wanting," "in perfection," or "in full operation."

141

Nothing's the matter, nothing's the matter with me;
That there's nothing the matter—this is the
  "Namu-amida-butsu."

142

To be grateful is not *anjin*;[27]
Nothing happening is nothing happening.
To be grateful is a fraud—[28]
'Tis true, 'tis true!

143

Whether I'm falling [to hell]
Or bound for the Pure Land—
I have no knowledge:
All is left to Amida's Vow.
"Namu-amida-butsu!"

144

Doubts have been taken away—
I know not how and when!
How to be thankful for the favor—I know not!
"Namu-amida-butsu!"

145

I am happy!
The root of sinfulness[29] is cut off;

---

[27] *Anjin*, literally, "mind pacified," meaning "faith confirmed."

[28] Literally, *bakemono* is "something unreal," "something temporarily assuming a certain shape but not at all genuine."

[29] *Bombu* in Japanese. Saichi uses the term also in an abstract sense, in the sense of *bombu*-hood, making it contrast with Buddhahood. Sinfulness here is not to be understood in its Christian sense.

From Saichi's Journals

Though still functioning, it is the same as
 non-existent.
How happy I am!
Born of happiness is the "Namu-amida-butsu."

146

"O Saichi, won't you tell us about *Tariki*?"
"Yes, but there is neither *Tariki* nor *Firiki*,
What is, is the graceful acceptance only."

147

Where are Saichi's evil desires gone?
They are still here:
I hate, I love, I crave—
How wretched, how wretched I am!

Epilogue

What World Perspectives Means
by Ruth Nanda Anshen

This is a reprint of Volume XII of the WORLD PERSPECTIVES SERIES, which the present writer has planned and edited in collaboration with a Board of Editors consisting of NIELS BOHR, RICHARD COURANT, HU SHIH, ERNEST JACKH, ROBERT M. MACIVER, JACQUES MARITAIN, J. ROBERT OPPENHEIMER, I. I. RABI, SARVEPALLI RADHAKRISHNAN, ALEXANDER SACHS.

This volume is part of a plan to present short books in a variety of fields by the most responsible of contemporary thinkers. The purpose is to reveal basic new trends in modern civilization, to interpret the creative forces at work in the East as well as in the West, and to point to the new consciousness which can contribute to a deeper understanding of the interrelation of man and the universe, the individual and society, and of the values shared by all people. *World Perspectives* represents the world community of ideas in a universe of discourse, emphasizing the principle of unity in mankind of permanence within change.

Recent developments in many fields of thought have opened unsuspected prospects for a deeper understanding of man's situation and for a proper appreciation of human values and human aspirations. These prospects, though the outcome of purely specialized studies in limited fields, require for their analysis and synthesis a new structure and frame in which they can be explored, enriched and advanced in all their aspects for the benefit of man and society. Such a structure and frame it is the endeavor of *World Perspectives* to define, leading hopefully to a doctrine of man.

A further purpose of this Series is to attempt to overcome a principal ailment of humanity, namely, the effects of the atomization of knowledge produced by the overwhelming accretion of facts which science has created, to clarify and synthesize ideas through the *depth* fertilization of minds; to show from diverse and important points of view the correlation of ideas, facts and values which are in perpetual interplay; to demonstrate the character, kinship, logic and operation of the entire organism of reality while showing the persistent interrelationship of the processes of the human mind and in the interstices of knowledge; to reveal the inner synthesis and organic unity of life itself.

It is the thesis of *World Perspectives* that in spite of the difference and diversity of the disciplines represented, there exists a strong common agreement among the authors concerning the overwhelming need for counterbalancing the multitude of compelling scientific activities and investigations of objective phenomena from physics to metaphysics, history and biology and to relate these to meaningful experience. To provide this balance, it is necessary to stimulate an awareness of the basic fact that ultimately the individual human personality must tie all the loose ends together into an organic whole, must relate himself to himself, to mankind and so-

What World Perspectives Means

ciety while deepening and enhancing his communion with the universe. To anchor this spirit and to impress it on the intellectual and spiritual life of humanity, on thinkers and doers alike, is indeed an enormous challenge which cannot be left entirely either to natural science on the one hand nor to organized religion on the other. For we are confronted with the unbending necessity to discover a principle of differentiation yet relatedness lucid enough to justify and purify scientific, philosophic and all other knowledge while accepting their mutual interdependence. This is the crisis in consciousness made articulate through the crisis in science. This is the new awakening.

World Perspectives is dedicated to the task of showing that basic theoretical knowledge is related to the dynamic content of the wholeness of life. It is dedicated to the new synthesis at once cognitive and intuitive. It is concerned with the unity and continuity of knowledge in relation to man's nature and his understanding, a task for the synthetic imagination and its unifying vistas. Man's situation is new and his response must be new. For the nature of man is knowable in many different ways and all of these paths of knowledge are interconnectable and some are interconnected, like a great network, a great network of people, between ideas, between systems of knowledge, a rationalized kind of structure which is human culture and human society.

Knowledge, it is shown in these volumes, no longer consists in a manipulation of man and nature as opposite forces, nor in the reduction of data to statistical order, but is a means of liberating mankind from the destructive power of fear, pointing the way toward the goal of the rehabilitation of the human will and the rebirth of faith and confidence in the human person. The works published also endeavor to reveal that the cry for patterns, systems and authorities is growing less insistent as the desire grows stronger in both East and West for the recovery of a dignity, integrity and self-realization which are the inalienable rights of man who is not a mere *tabula rosa* on which anything may be arbitrarily imprinted by external circumstance but who possesses the unique potentiality of free creativity. Man is differentiated from other forms of life in that he may guide change by means of conscious purpose in the light of rational experience.

World Perspectives is planned to gain insight into the meaning of man who not only is determined by history but who also determines history. History is to be understood as concerned not only with the life of man on this planet but as including also such cosmic influences as interpenetrate our human world. This generation is discovering that history does not conform to the social optimism of modern civilization and that the organization of human communities and the establishment of freedom, justice and peace are not only intellectual achievements but spiritual and moral achievements as well, demanding a cherishing of the wholeness of human personality, the "unmediated wholeness of feeling and thought," and constituting a never-ending challenge to man, emerging from the abyss of meaninglessness and suffering, to be renewed and replenished in the totality of his life.

What World Perspectives Means

World Perspectives is committed to the recognition that all great changes are preceded by a vigorous intellectual reevaluation and reorganization. Our authors are aware that the sin of hybris may be avoided by showing that the creative process itself is not a free activity if by free we mean arbitrary or unrelated to cosmic law. For the creative process in the human mind, the developmental process in organic nature and the basic laws of the inorganic realm may be but varied expressions of a universal formative process. Thus *World Perspectives* hopes to show that although the present apocalyptic period is one of exceptional tensions, there is also an exceptional movement at work toward a compensating unity which cannot obliterate the ultimate moral power pervading the universe, that very power on which all human effort must at last depend. In this way, we may come to understand that there exists an independence of spiritual and mental growth which though conditioned by circumstances is never determined by circumstances. In this way the great plethora of human knowledge may be correlated with an insight into the nature of human nature by being attuned to the wide and deep range of human thought and human experience. For what is lacking is not the knowledge of the structure of the universe but a consciousness of the qualitative uniqueness of human life.

And finally, it is the thesis of this Series that man is in the process of developing a new awareness which, in spite of his apparent spiritual and moral captivity, can eventually lift the human race above and beyond the fear, ignorance, brutality and isolation which beset it today. It is to this nascent consciousness, to this concept of man born out of a fresh vision of reality, that *World Perspectives* is dedicated.

BOARD OF EDITORS

of

WORLD PERSPECTIVES

Sir Kenneth Clark
Richard Courant
Werner Heisenberg
Ivan Illich
Konrad Lorenz
Robert M. MacIver
Joseph Needham
I. I. Rabi
Sarvepalli Radhakrishnan
Karl Rahner, S.J.
Alexander Sachs
C. N. Yang

WORLD PERSPECTIVES

Volumes already published

I   APPROACHES TO GOD — Jacques Maritain
II ACCENT ON FORM — Lancelot Law Whyte
III SCOPE OF TOTAL ARCHITECTURE — Walter Gropius
IV RECOVERY OF FAITH — Sarvepalli Radhakrishnan
V   WORLD INDIVISIBLE — Konrad Adenauer
VI SOCIETY AND KNOWLEDGE — V. Gordon Childe
VII THE TRANSFORMATIONS OF MAN — Lewis Mumford
VIII MAN AND MATERIALISM — Fred Hoyle
IX THE ART OF LOVING — Erich Fromm
X   DYNAMICS OF FAITH — Paul Tillich
XI MATTER, MIND AND MAN — Edmund W. Sinnott
XII MYSTICISM: CHRISTIAN AND BUDDHIST — Daisetz Teitaro Suzuki
XIII MAN'S WESTERN QUEST — Denis de Rougemont
XIV AMERICAN HUMANISM — Howard Mumford Jones
XV THE MEETING OF LOVE AND KNOWLEDGE — Martin C. D'Arcy, S.J.
XVI RICH LANDS AND POOR — Gunnar Myrdal
XVII HINDUISM: ITS MEANING FOR THE LIBERATION OF THE SPIRIT — Swami Nikhilananda

XVIII	Can People Learn to Learn?	Brock Chisholm
XIX	Physics and Philosophy	Werner Heisenberg
XX	Art and Reality	Joyce Cary
XXI	Sigmund Freud's Mission	Erich Fromm
XXII	Mirage of Health	René Dubos
XXIII	Issues of Freedom	Herbert J. Muller
XXIV	Humanism	Moses Hadas
XXV	Life: Its Dimensions and Its Bounds	Robert M. MacIver
XXVI	Challenge of Psychical Research	Gardner Murphy
XXVII	Alfred North Whitehead: His Reflections on Man and Nature	Ruth Nanda Anshen
XXVIII	The Age of Nationalism	Hans Kohn
XXIX	Voices of Man	Mario Pei
XXX	New Paths in Biology	Adolf Portmann
XXXI	Myth and Reality	Mircea Eliade
XXXII	History as Art and as Science	H. Stuart Hughes
XXXIII	Realism in Our Time	Georg Lukács
XXXIV	The Meaning of the Twentieth Century	Kenneth E. Boulding
XXXV	On Economic Knowledge	Adolph Lowe
XXXVI	Caliban Reborn	Wilfrid Mellers
XXXVII	Through the Vanishing Point	Marshall McLuhan and Harley Parker
XXXVIII	The Revolution of Hope	Erich Fromm
XXXIX	Emergency Exit	Ignazio Silone
XL	Marxism and the Existentialists	Raymond Aron
XLI	Physical Control of the Mind	José M. R. Delgado, M.D.